DOWN TO EARTH

Geographic Information for Sustainable Development in Africa

Committee on the Geographic Foundation for Agenda 21
Committee on Geography
Mapping Science Committee

Board on Earth Sciences and Resources

Division on Earth and Life Studies

NATIONAL RESEARCH COUNCIL
OF THE NATIONAL ACADEMIES

THE NATIONAL ACADEMIES PRESS
Washington, D.C.
www.nap.edu

THE NATIONAL ACADEMIES PRESS • 500 Fifth Street, N.W. • Washington, DC 20001

NOTICE: The project that is the subject of this report was approved by the Governing Board of the National Research Council, whose members are drawn from the councils of the National Academy of Sciences, the National Academy of Engineering, and the Institute of Medicine. The members of the committee responsible for the report were chosen for their special competences and with regard for appropriate balance.

This study was jointly sponsored by the Environmental Systems Research Institute, National Aeronautics and Space Administration, National Imagery and Mapping Agency, National Oceanic and Atmospheric Administration, National Science Foundation, U.S. Agency for International Development, U.S. Department of State, and U.S. Geological Survey. Any opinions, findings, conclusions, or recommendations expressed in this publication are those of the authors and do not necessarily reflect the view of the organizations or agencies that provided support for this project.

International Standard Book Number: 0-309-08478-4
Library of Congress Control Number: 2002110505

Front cover: *Left*—a GeoCover-Ortho image of Mount Kilimanjaro originally obtained at 30 × 30 m spatial resolution. It has a positional accuracy of better than 50 m (root mean square error). Landsat TM bands 7,4,2 (mid-infrared, near-infrared, and green) are displayed (courtesy of Earth Satellite Corporation). Each color or shade is unique and depends on the vegetation type, health, and growth stage. The bright greens are dense vegetation. The purples and pinks are sparse to no vegetation. The bottom third center of the image along Mount Kilimanjaro's lower slopes contains areas of clear cuts (in pinks) surrounded by uncut verdant forest (bright greens). The top of the mountain is snow-covered (blue) and the white areas are clouds. *Upper Right*—artist's rendition of the Shuttle Radar Topography Mission 60-m (200-ft) mast being deployed from the space shuttle *Endeavor* (courtesy NASA Jet Propulsion Laboratory). Radar images are collected from the end of the mast and from the shuttle payload bay. *Lower Right*—paper maps used in decision support in Namibia (courtesy of Jo Tagg, Namibia Nature Foundation).

Cover designed by Van Nguyen

Copies of this report are available from:

The National Academies Press
500 Fifth Street, N.W.
Washington, DC 20055
Lockbox 285
800-624-6242
202-334-3313 (in the Washington metropolitan area)
http://www.nap.edu

Printed in the United States of America

THE NATIONAL ACADEMIES

Advisers to the Nation on Science, Engineering, and Medicine

The **National Academy of Sciences** is a private, nonprofit, self-perpetuating society of distinguished scholars engaged in scientific and engineering research, dedicated to the furtherance of science and technology and to their use for the general welfare. Upon the authority of the charter granted to it by the Congress in 1863, the Academy has a mandate that requires it to advise the federal government on scientific and technical matters. Dr. Bruce M. Alberts is president of the National Academy of Sciences.

The **National Academy of Engineering** was established in 1964, under the charter of the National Academy of Sciences, as a parallel organization of outstanding engineers. It is autonomous in its administration and in the selection of its members, sharing with the National Academy of Sciences the responsibility for advising the federal government. The National Academy of Engineering also sponsors engineering programs aimed at meeting national needs, encourages education and research, and recognizes the superior achievements of engineers. Dr. Wm. A. Wulf is president of the National Academy of Engineering.

The **Institute of Medicine** was established in 1970 by the National Academy of Sciences to secure the services of eminent members of appropriate professions in the examination of policy matters pertaining to the health of the public. The Institute acts under the responsibility given to the National Academy of Sciences by its congressional charter to be an adviser to the federal government and, upon its own initiative, to identify issues of medical care, research, and education. Dr. Harvey V. Fineberg is president of the Institute of Medicine.

The **National Research Council** was organized by the National Academy of Sciences in 1916 to associate the broad community of science and technology with the Academy's purposes of furthering knowledge and advising the federal government. Functioning in accordance with general policies determined by the Academy, the Council has become the principal operating agency of both the National Academy of Sciences and the National Academy of Engineering in providing services to the government, the public, and the scientific and engineering communities. The Council is administered jointly by both Academies and the Institute of Medicine. Dr. Bruce M. Alberts and Dr. Wm. A. Wulf are chair and vice chair, respectively, of the National Research Council.

www.national-academies.org

COMMITTEE ON THE GEOGRAPHIC FOUNDATION FOR AGENDA 21

JOHN R. JENSEN, *Chair*, University of South Carolina, Columbia
KWESI BOTCHWEY, Harvard University, Cambridge, Massachusetts
ELLEN BRENNAN-GALVIN, Woodrow Wilson International Center for Scholars, Washington, D.C.
CHRISTIAN J. JOHANNSEN, Purdue University, West Lafayette, Indiana
CALESTOUS JUMA, Harvard University, Cambridge, Massachusetts
AKINLAWON L. MABOGUNJE, Development Policy Center, Ibadan, Nigeria
ROBERTA BALSTAD MILLER, Columbia University, Palisades, New York
KEVIN P. PRICE, University of Kansas, Lawrence
PRISCILLA A. C. REINING *(Retired)*, International Office of the American Association for the Advancement of Science, Washington, D.C.
DAVID L. SKOLE, Michigan State University, East Lansing
ANDREW STANCIOFF, Stone Environmental, Inc., Washington, D.C.
D. R. FRASER TAYLOR, Carleton University, Ottawa, Ontario, Canada

NRC Staff

ANTHONY R. de SOUZA, Director, Board on Earth Sciences and Resources
PAUL M. CUTLER, Study Director
LISA M. VANDEMARK, Study Director
KRISTEN L. KRAPF, Program Officer
EILEEN M. McTAGUE, Research Assistant
TERESIA K. WILMORE, Project Assistant

COMMITTEE ON GEOGRAPHY

BILLIE L. TURNER II, *Chair*, Clark University, Worcester, Massachusetts
BERNARD O. BAUER, University of Southern California, Los Angeles
RUTH S. DEFRIES, University of Maryland, College Park
ROGER M. DOWNS, Pennsylvania State University, University Park
MICHAEL F. GOODCHILD, University of California, Santa Barbara
SUSAN HANSON, Clark University, Worcester, Massachusetts
SARA L. MCLAFFERTY, University of Illinois, Urbana
ELLEN S. MOSLEY-THOMPSON, The Ohio State University, Columbus
ERIC S. SHEPPARD, University of Minnesota, Minneapolis

NRC Staff

KRISTEN L. KRAPF, Program Officer
MONICA R. LIPSCOMB, Research Assistant
VERNA J. BOWEN, Administrative Associate

MAPPING SCIENCE COMMITTEE

DAVID J. COWEN, *Chair*, University of South Carolina, Columbia
ANNETTE J. KRYGIEL, *Vice-Chair*, Independent Consultant, Integro, Great Falls, Virginia
ERIC A. ANDERSON, City of Des Moines, Iowa
WILLIAM J. CRAIG, University of Minnesota, Minneapolis
MARK MONMONIER, Syracuse University, New York
JOEL MORRISON, Ohio State University, Columbus
SHERYL G. OLIVER, Illinois Department of Natural Resources, Springfield
HARLAN J. ONSRUD, University of Maine, Orono
C. STEPHEN SMYTH, MobileGIS, Ltd., Bellevue, Washington
JAMES V. TARANIK, University of Nevada, Reno
REX W. TRACY, BAE Systems, San Diego, California
A. KEITH TURNER, Colorado School of Mines, Golden

NRC Staff

PAUL M. CUTLER, Program Officer
RADHIKA S. CHARI, Senior Project Assistant

BOARD ON EARTH SCIENCES AND RESOURCES

RAYMOND JEANLOZ, *Chair*, University of California, Berkeley
JILL BANFIELD, University of California, Berkeley
STEVEN R. BOHLEN, Joint Oceanographic Institutions, Washington, D.C.
VICKI J. COWART, Colorado Geological Survey, Denver
DAVID L. DILCHER, University of Florida, Gainesville
ADAM M. DZIEWONSKI, Harvard University, Cambridge, Massachusetts
WILLIAM L. GRAF, University of South Carolina, Columbia
RHEA GRAHAM, New Mexico Interstate Stream Commission, Albuquerque
GEORGE M. HORNBERGER, University of Virginia, Charlottesville
DIANNE R. NIELSON, Utah Department of Environmental Quality, Salt Lake City
MARK SCHAEFER, NatureServe, Arlington, Virginia
BILLIE L. TURNER II, Clark University, Worcester, Massachusetts
THOMAS J. WILBANKS, Oak Ridge National Laboratory, Tennessee

NRC Staff

ANTHONY R. de SOUZA, Director
TAMARA L. DICKINSON, Senior Program Officer
DAVID A. FEARY, Senior Program Officer
ANNE M. LINN, Senior Program Officer
PAUL M. CUTLER, Program Officer
KRISTEN L. KRAPF, Program Officer
KERI H. MOORE, Program Officer
LISA M. VANDEMARK, Program Officer
YVONNE P. FORSBERGH, Research Assistant
MONICA R. LIPSCOMB, Research Assistant
EILEEN M. McTAGUE, Research Assistant
JENNIFER T. ESTEP, Administrative Associate
VERNA J. BOWEN, Administrative Associate
RADHIKA S. CHARI, Senior Project Assistant
KAREN L. IMHOF, Senior Project Assistant
SHANNON L. RUDDY, Senior Project Assistant
TERESIA K. WILMORE, Project Assistant
WINFIELD SWANSON, Editorial Consultant

Acknowledgments

This report has been reviewed in draft form by individuals chosen for their diverse perspectives and technical expertise, in accordance with procedures approved by the National Research Council's Report Review Committee. The purpose of this independent review is to provide candid and critical comments that will assist the institution in making its published report as sound as possible and to ensure that the report meets institutional standards for objectivity, evidence, and responsiveness to the study charge. The review comments and draft manuscript remain confidential to protect the integrity of the deliberative process. We wish to thank the following individuals for their review of this report:

Dr. André Bassolé
Environmental Information Systems in sub-Saharan Africa (EIS-AFRICA)
Ouagadougou, Burkina Faso

Dr. Ruth S. Defries
University of Maryland
College Park

Dr. Paul V. Desanker
University of Virginia
Charlottesville

Dr. Michael F. Goodchild
University of California
Santa Barbara

Dr. James Guseh
North Carolina Central University
Durham

Dr. Barry N. Haack
George Mason University
Fairfax, Virginia

Dr. David Kaplan
Department of Trade and Industry
Cape Town, South Africa

Dr. Pamela A. Matson
Stanford University
California

Dr. John Mugabe
African Centre for Technology Studies
Nairobi, Kenya

Dr. Marilyn Silberfein
Temple University
Philadelphia, Pennsylvania

Although the reviewers listed above have provided many constructive comments and suggestions, they were not asked to endorse the conclusions or recommendations nor did they see the final draft of the report before its release. The review of this report was overseen by Dr. Thomas J. Wilbanks, Oak Ridge National Laboratory, and Dr. Brian J. L. Berry, University of Texas. Appointed by the National Research Council, they were responsible for making certain that an independent examination of this report was carried out in accordance with institutional procedures and that all review comments were carefully considered. Responsibility for the final content of this report rests entirely with the authoring committee and the institution.

Preface

On July 9, 2001, Undersecretary of State for Global Affairs Paula Dobriansky sent a letter to Dr. Bruce Alberts, president of the U.S. National Academy of Sciences, requesting a study as a contribution to the U.S. Department of State's "Geographic Information for Sustainable Development" Alliance for the World Summit on Sustainable Development in Johannesburg in August 2002. Being held a decade after the United Nations Conference on Environment and Development in Rio de Janeiro, the main goals of the summit are to "reinvigorate the global commitments to and achieve a higher level of international solidarity and partnership in the promotion of sustainable development" (UN, 2001).

The Geographic Information for Sustainable Development Alliance is an international collaboration and alliance whose objective is to apply a new generation of earth observation data and GIS-linked technologies to ongoing sustainable development problems in Africa. The alliance focuses on four case-study regions in sub-Saharan Africa. These are the Upper Niger basin, the Kenya-Tanzania coast, the African Great Lakes Region, and the Limpopo and Zambezi river basins. As a component of the Geographic Information for Sustainable Development Alliance, this study concentrates on sub-Saharan Africa and draws on experiences from activities in these case-study regions. Descriptions of ongoing activities in these areas are provided as examples of the application of geographic information to sustainable development in Africa. Given the embryonic state of some activities in the case-study regions and the available time and resources, the committee chose not to critically analyze these efforts. Instead the committee (Appendix A) drew on literature and testimony from public, private, and non-profit organizations working with geographic information and applications in Africa (Appendix B) and its own experience and judgment to determine broad lessons learned. Committee and staff members also participated in international conferences and meetings of geographic information practitioners in Bamako, Mali; Dar es Salaam, Tanzania; Nairobi, Kenya; Niamey, Niger; and Ouagadougou, Burkina Faso.

In a symposium at the U.S. National Academy of Sciences in 1999 Professor John E. Estes first noted the need for a study of this type. He suggested compiling a resource highlighting the value of geographic data and tools for addressing issues of sustainable development. Professor Estes stated, "We cannot have sustainable economic development and improved environmental quality without understanding how our global resource base is changing through time."

In addition to the U.S. Department of State the study received support from the Environmental Systems Research Institute, National Aeronautics and Space Administration, National Imagery and Mapping Agency, National Oceanic and Atmospheric Administration, National Science Foundation, U.S. Agency for International Development, and the U.S. Geological Survey.

John R. Jensen, Chair

REFERENCE

UN (United Nations). 2001. SADC Progress Report on the Implementation of Agenda 21 and Sustainable Development: A Report to the 2002 World Summit on Sustainable Development. Available at <http://www.johannesburgsummit.org/web_pages/sadc_prepcom_progress_report.pdf>. Accessed August 1, 2002.

Contents

Executive Summary

In 1992 world leaders adopted Agenda 21 (UNCED, 1992), the work program of the 1992 U.N. Conference on Environment and Development in Rio de Janeiro. The landmark event provided a political foundation and action items to facilitate the global transition toward sustainable development (Box ES-1). The international community is marking the tenth anniversary of this conference by holding the World Summit on Sustainable Development in Johannesburg, South Africa, in August 2002. The goals of the summit are to "reinvigorate the global commitments to, and achieve a higher level of international solidarity and partnership in the promotion of sustainable development" (UN, 2001). The summit builds on the political momentum created by the U.N. Millennium Declaration, in which world leaders committed themselves to achieving a broad range of time-bound international sustainable development objectives for which sustainable development provides a unifying framework (UN, 2001).

This report is a component of the U.S. State Department's contribution to the World Summit on Sustainable Development, the "Geographic Information for Sustainable Development" project (GISD, 2002). Because South Africa is hosting the summit, it seemed appropriate for this report to focus on sub-Saharan Africa. Examples are drawn from case-study regions where the U.S. Agency for International Development (USAID) and other agencies have broad experience. Although African countries are the geographic focus of the report, the material in the report has broader applicability.

PURPOSE OF THIS REPORT

This report summarizes the importance and applicability of geographic[1] data for sustainable development. Geographic data describe spatial variations across the landscape at a variety of scales (local, national, global) and include such elements as climate, elevation, soil, vegetation, population, land use, and economic activity. The report draws on experiences in African countries and examines how future sources and applications of geographic data could provide reliable support to decision-makers[2] as they work toward sustainable development. The committee emphasizes the potential of new technologies, such as satellite remote-sensing systems[3] and geographic information systems (GIS), that have revolutionized data collection and analysis over the last decade.

The Charge to the Committee

As a component of the State Department's contribution to the World Summit in Johannesburg (August 2002), this study will examine the geographic foundation for natural resource management and development issues in Africa. Centered on a place-based,[4] integrative framework, the study draws on experiences of U.S. government agencies, international groups, decision-makers, and experts to examine

(1) existing remote-sensing and GIS efforts in case-study regions and lessons learned from those efforts;

---

[1]*Geography* is an integrative discipline that brings together the physical and human dimensions of the world in the study of people, places, and environments. Its subject matter is the Earth's surface and the processes that shape it, the relationships between people and environments, and the connections between people and places (Geography Education Standards Project, 1994).

[2]*Decision-makers* in the context of sustainable development choose actions that directly or indirectly affect the environment and reside in all levels of government, the citizenry, the private sector, non-governmental organizations, or overseas development agencies, for example.

[3]*Remote-sensing* is "the measurement or acquisition of information of some property of an object or phenomenon by a recording device that is not in physical or intimate contact with the object or phenomenon under study" (Colwell, 1997). Satellites and aircraft are common platforms for remote-sensing systems.

[4]*Place-based* studies are "the systematic analysis of social, economic, political, and environmental processes operating in a place that provides an integrated understanding of its distinctiveness or character" (NRC, 1999).

BOX ES-1
Sustainable Development: A Matter of Definition

The concept of sustainable development has a long history in scientific thought. As early as 1749 the Swedish botanist Linnaeus in his *Oeconomia naturae* linked economy to nature in a way that resembles many of the concepts of sustainable development. His economic program focused on the need to make efficient use of existing resources rather than pursue military expeditions as a means of economic survival. Over 200 years later in *Our Common Future* the World Commission on Environment and Development (WCED, 1987) gave international prominence to sustainable development and defined it as "development that meets the needs of the present without compromising the ability of future generations to meet their own needs." *Our Common Journey*, a report of the U.S. National Research Council (NRC, 1999), added a temporal dimension, defining sustainable development as "the reconciliation of society's developmental goals with its environmental limits over the long term." These definitions reflect the growing need to provide an ethical framework for integrating developmental and environmental goals. Sustainable development is therefore a set of guiding principles whose implementation is reflected in a variety of action programs, of which Agenda 21 is the most prominent.

(2) existing levels of local expertise and technology, and ongoing efforts in geospatial capacity building[5];
(3) a range of questions relating to the practical application of new and existing spatial data (e.g., required resolution, challenges of integrating layers of environmental and social data, and baseline data against which future change can be measured);
(4) the role of decision-support systems in the application of these data; and
(5) options for making efforts sustainable beyond 2002.

The geographic foundation for Agenda 21 in Africa involves a wide array of geographic data, tools, and perspectives (including social, environmental, and economic data; maps and models; and the analysis of pattern and processes, place, and scale). This study was undertaken in support of the U.S. Geographic Information for Sustainable Development Alliance, which focuses on the uses of Earth observation data and GIS to address Agenda 21 issues. In this context the report emphasizes fundamental data types that are needed in many applications and ways of increasing accessibility to these data in Africa. An assessment of the potential applications of the full range of geographic data, tools, and concepts for natural resource management and development in Africa would also be valuable.

GEOGRAPHIC DATA AND SUSTAINABLE DEVELOPMENT

Chapter 40 in Agenda 21, "Information[6] for Decision-Making," stresses the need for more and different types of data to be collected at all scales to track the status and trends of Earth's ecosystems, natural resources, pollution, and socioeconomic variables. Chapter 40 concludes that "the gap in the availability, quality, coherence, standardization and accessibility of data between the developed[7] and the developing world has been increasing, seriously impairing the capacities of countries to make informed decisions concerning environment and development."[8]

Five years after the U.N. Conference on Environment and Development (UNCED), in a 1997 assessment of the state of data supporting decision-making on each of Agenda 21 action items (UN, 1997), most of the nine responding African governments[9] described their existing databases as "poor" or containing "some good data but many gaps." Ten years after UNCED, the work of the U.N. Economic Commission for Africa (ECA), which presses for greater awareness of the significance of geographic information in socioeconomic development among African governments and other sectors[10] (ECA, 2001), illustrates that much remains to be achieved in applying geographic information to sustainable development.

Geographic data are obtained from ground-based (in situ) measurements or from remote-sensing systems. These data are of little practical value in sustainable development decision-making if they cannot be analyzed in conjunction with development data, such as economic or health data, that are geographically referenced[11] (Jensen, 2000). Data that de-

[5]*Capacity* is the ability to undertake certain activities, solve problems and achieve objectives (Fukuda-Parr et al., 2002) such as interpreting a map (an example of human capacity), possessing a computer (organizational capacity), or sharing data (societal capacity). Capacity is built by enhancing these abilities.

[6]*Information* is data that humans assimilate and evaluate to solve a problem or make a decision (EIS-Africa, 2001).

[7]Developed countries are those with "high" gross national income (World Bank, 2002). Developing countries are those with "low" or "medium" gross national income. The committee adopted these terms in the report.

[8]*Development* includes at least four related concepts (Dernbach, 2002): peace and security, economic development, social development, and national governance that secures peace and development.

[9]Algeria, Benin, Egypt, Gabon, Malawi, South Africa, Tunisia, Uganda, Zimbabwe

scribe environment and development can be linked by geographic location to provide greater understanding of complex issues, and GISs were developed specifically for this purpose. A GIS is formed from a set of map layers or overlays registered to a common geographic coordinate system and is "a digital information system that is designed to work with data referenced by spatial or geographic coordinates" (Star and Estes, 1990). Geographic information systems are powerful tools for ingesting, storing, retrieving, transforming, processing, and displaying geographic data (Burrough, 1986). If a GIS involves the integration of geographically referenced data in a problem-solving situation, it can become important for decision-making, or a "decision-support system" (Cowen, 1988). Geospatial capacity is essential for all these steps so that full use can be made of the capabilities of geographic data for supporting sustainable-development decision-making.

The United States provides geographic data for a wide variety of applications in Africa. These data include:

- data from the 24-satellite Global Positioning System (GPS) (Chapter 5);
- global 30 × 30 m orthorectified Landsat Thematic imagery from circa 1990 through the National Aeronautics and Space Administration (NASA) Data Buy and Earth Satellite Corporation (Chapter 6);
- imagery of many African countries from CORONA data and Space Shuttle photography;
- global digital elevation model information (90 × 90 m spatial resolution) derived from NASA's Shuttle Radar Topography Mission (Chapter 5) (in processing);
- hydrologic information derived from the Global Topography at 30 arc seconds (GTOPO30) dataset (Chapter 5);
- land cover derived from satellite sensors including Landsat, NASA's Terra Moderate Resolution Imaging Spectroradiometer (MODIS), and the National Oceanic and Atmospheric Administration's (NOAA's) Advanced Very High Resolution Radiometer (AVHRR) (Chapter 6);
- remote-sensing-derived vegetation indexes, including the Normalized Difference Vegetation Index (NDVI) (Chapter 6);
- rainfall measurements from the Tropical Rainfall Measuring Mission (Chapter 6);
- soil moisture measurements from the Defense Meteorological Satellite Program (DMSP);
- estimation of human population distribution using LandScan 2000 and Gridded Population of the World datasets (Chapter 5); and
- fire monitoring using DMSP nighttime lights and Terra MODIS imagery (Chapter 6).

Geographic data and information such as these are fundamental for addressing Agenda 21 issues. For example, they are used in early warning systems for natural disasters, human and livestock health, and crop production, and for monitoring soil erosion, rainfall, crop yields, disease vectors (e.g., insects), and biodiversity. Decision-support systems that include geographic information and tools (e.g., GIS) are used in decisions about land use, water-resource allocation, flood and erosion prevention, and natural resource management. Additionally, decision-support systems in the form of two-dimensional maps and images promote the application of geographic information at the local level where many sustainable development challenges occur (e.g., Chapters 3 and 7).

The next section discusses lessons learned about the application of geographic information in the four case-study regions and other areas in Africa. The committee's conclusions and recommendations are presented in the final section. The conclusions and recommendations follow the three-section structure of the report: namely, the spatial data and telecommunications infrastructure (Chapter 4), geographic data and tools (Chapters 5, 6, and 7), and geospatial capacity building (Chapter 8).

LESSONS LEARNED

Africa has a small but growing community of geographic data providers, processors and analysts, trainers, technicians, advocates, and data and information users (decision-makers). The community's growth is demonstrated by the more than 400 participants at the Africa-GIS conference in Nairobi in November 2001 compared to the 70 attendees at the first Africa-GIS conference in Tunis in 1993.[12] As the community grows, its activities are becoming better coordinated. This community comprises African and international partners from non-governmental organizations (NGOs), universities, private companies, and foreign governments, including the space and aid agencies that are a major source of geographic data, training, and support.

Efforts to expand the use of geographic information at national and regional levels are resulting in data and information for decision-making, technical training for students and professionals, and creation of geospatial capacity. The capacity to manage and use geographic data and information is growing through continent-wide activities (e.g., EIS-Africa and ECA's regional centers) and partnerships (e.g., NOAA's Radio and Internet for the Communication of Hydro-Meteorological and Climate-related Information across Africa [RANET] project, the Miombo Network, and the

[10]The committee applies the term "sector" to groupings such as environmental, social, and economic sectors.

[11]These are data with known latitude, longitude, and elevation, or other horizontal and vertical coordinates.

[12]Daniel Tunstall, World Resources Institute, personal communication, May 10, 2002.

Food and Agricultural Organization's [FAO's] Africover project). Some of these activities like the Famine Early Warning System Network (FEWS NET) have been in place for many years, whereas others like Uganda's Advocates Coalition for Development and Environment (ACODE), Burkina Faso's National Program for Environment Information Management (PNGIM), and the Livestock Early Warning System (LEWS) in East Africa are new.

Needs-Driven Approaches and Data Sharing

Needs-driven approaches and open data-sharing environments are common among effective applications of geographic information (e.g., the Community Based Natural Resources Management [CBNRM] program in Namibia, the continent-wide Mapping Malaria Risk in Africa [MARA] project, and the Miombo Network in southern Africa). The needs-driven approach of the CBNRM program has built credibility with field users, led to a strong feeling of ownership by rural people and field-based support staff, fostered a culture of sensitivity to community needs among technical institutions that are partners in the program, generated trust and a common vision among partners (communities, government, donors), and built a critical mass to enhance sustainability of the program. Data sharing, facilitated by all users adopting the same software, data formats, and file directory structure, and a metadata database has resulted in cost savings.

The agricultural and natural resource management sectors are a likely primary source of demand for geographic information and related decision-support tools. These sectors are the main users because the livelihoods of the majority of Africans depend on them. Additional demand will arise as African countries need to satisfy reporting requirements on treaties to which they are signatories.

Lesson Learned: Needs-driven as opposed to prescriptive approaches with provision of information in appropriate and usable forms are most likely to result in effective application of geographic information.

From Environmental Management to Sustainable Development

Geographic information and technologies are central to achieving a successful transition from traditional environmental and resource management practices to sustainable development because of their integrative quality (linking social, economic, and environmental data) and their place-based quality (addressing relationships among places at local, national, regional, and global scales).

A narrow focus on either economic development or environmental management can obscure the connections between environmental change and social, political, and economic activities, artificially separating environment from development. This separation can result in short-term, project-oriented data collection; single-issue development agendas (e.g., economic growth divorced from environmental and intergenerational equity considerations); and spurious attempts to make tradeoffs between inseparable dimensions of sustainable development such as human well-being and environmental protection (NRC, 2002).

Sustainable development necessarily links people, their needs, and the impacts of their behavior over time (including patterns of population growth and consumption, cultural patterns, and political activities) to the environment and the economy. Consequently, data on human population distribution are fundamentally important to decision-makers as they address Agenda 21 issues.

Lesson Learned: Geographic information and technologies are central to the transition from traditional environmental management to sustainable development that brings people to the fore, rightfully integrating environment and development.

Geographic Information at the Intersection of Sectors

Agenda 21 (UNCED, 1992) calls for integrated social, economic, and environmental data. There is growing recognition by decision-makers in Africa that problems at the intersection of agriculture and environmental management, climate change, and land-cover change, with their attendant social and economic consequences, will be at the forefront in the new century.

Technological advances fostering the integration of satellite imagery with other data (such as socioeconomic or health data) in GIS are opening new ways to synthesize complex and diverse geographic datasets, creating new opportunities for collaboration among natural and social scientists and decision-makers at all levels (e.g., the LEWS project, the Miombo Network, the MARA project, CBNRM, and Southern African Development Community [SADC]).

Lesson Learned: In this century many environmental problems will occur at the intersection of sectors. Geographic information technologies can assist people in tackling this integration challenge.

Good Governance

Societal capacity is built by governance[13] that promotes the relationships among individuals, organizations, and the larger society. In this way governance contributes to the de-

[13]Governance is defined by the UN Development Program (UNDP) as "the exercise of political, economic, and administrative authority to manage a nation's affairs." Sound or good governance is defined as that sub-set of governance, "wherein public resources and problems are managed efficiently and in response to the critical needs of society" (UNDP, 1997).

velopment of geospatial capacity. Linkages that facilitate collaboration among academics, governmental and non-governmental actors, and the private sector are needed for the transition to sustainable development (NRC, 1999).

Human and organizational capacity to apply geographic information and technology to Agenda 21 issues cannot grow or be maintained unless rooted in a wider societal context that values the contributions of science and technology, upholds principles of openness and sharing of information, and provides incentives for change and adaptation. The development of a policy environment that supports the use of geographic information depends on the attention given to scientific and technological issues in general.

Geographic data, hardware, and software systems are increasingly sophisticated but it is really the political, social, economic, and educational institutions of a country that ultimately determine the application and use of these data and tools for decision-making. Good governance creates a climate in which geospatial capacity can grow and vice versa. Geographic information illuminates social and political problems, such as the uneven distribution of the benefits of economic development, lack of accountability of elected officials, and a burden of disease that impacts societal cohesion.

Lesson Learned: Good governance promotes geospatial capacity and vice versa. Access to integrated geographic information allows civil society to hold government accountable; and government creates policies that determine public access to information and public participation in the decision process.

Barriers for Use of Geographic Information

There remain barriers to effective use of geographic information in Africa, including

- technical limitations of accessibility to data such as inadequate telecommunications infrastructure, limited bandwidth, and low Internet connectivity (Chapter 4);
- administrative challenges of accessibility to data, including lack of (1) familiarity on the part of government officials with requests for information, (2) efficient protocols for requesting government data (Chapter 8), (3) common data standards to promote sharing (Chapter 4), and (4) issues of copyright and distribution;
- inability to afford needed data and lack of availability of hard currency and foreign exchange in many countries (Chapter 6);
- educational and organizational limitations on access to data and technology including a poorly trained workforce and limited private-sector demand to spur the development of geographic information and tools (Chapter 8); and
- ineffective transfer of technology to the local level where many decisions are made that impact sustainable development (Chapter 7; NRC, 1999).

The available data often are not of sufficiently high spatial or temporal resolution to be useful for decision-support at the local level. Urban planners require regularly updated data at 1-m spatial resolution to take into account the rapid pace of change in cities. In rural areas where the bulk of the population still live the minimum spatial resolution of value to agricultural extension workers and rural development specialists is that of the small farms. Existing coverages, as outlined in this report, are impressive at national and sub-national levels but virtually nonexistent at local scales. The problem is confounded by the fact that what data are available rarely reach the rural and urban decision-makers at the local level dealing with the day-to-day realities of sustainable development.

In addition to data-availability challenges, many decision-makers in developed and developing countries have no experience with GIS and other spatial decision-support tools, and thus do not appreciate their potential for using geographic information. Other impediments to implementation of spatial decision-support systems include the orientation of projects toward data production rather than application, lack of planning for the decision-support process, lack of communication between technicians and scientists within an organization, and lack of inclusion of university research that could drive data analysis (EIS-Africa, 2001). With limited geographic data and a limited appreciation for its value the ability of African countries to address Agenda 21 issues and to fulfill their international treaty obligations for environmental reporting is compromised.

Lesson Learned: There are several barriers to the use of geographic data to address Agenda 21 issues. The next section describes approaches to overcome some of these barriers.

CONCLUSIONS AND RECOMMENDATIONS

Enabling Frameworks

Spatial Data Infrastructures

Conclusion: There is no universally accepted framework for geographic data management in Africa. An integrated, interoperable approach would provide Africans with better access to more diversified data that could then be applied to specific questions or problems. Decision-making on Agenda 21 issues requires access to data from multiple sources, including international ones, and this is facilitated by standardization within a spatial data infrastructure (SDI) (Chapter 4). Countries could benefit economically from SDIs because of the possibility to use data many times for many applications.

Recommendation: Because of the potential benefits, developing countries should consider using a standardized SDI that is compatible with the emerging Global Spatial Data Infrastructure (GSDI). Data derived from international development programs (for example, those of USAID) should conform to the standards recommended by the GSDI. In this way data collected by these programs are rendered more useful.

Telecommunications Infrastructure

Conclusion: Sustainable development activities would be improved if a greater emphasis were placed on distributed systems that enabled access to multiple geographic datasets and linked networks of African scientists, data users, and organizations. An efficient telecommunications infrastructure facilitates accessibility, use, and dissemination of geographic data and information, and forms the backbone of any SDI. Although telecommunications infrastructures are improving, in Africa as in much of the developing world they often are inadequate to support efficient SDIs (Chapter 4). Access to geographic data through the Internet is limited, and high connection cost and low bandwidth restrict data sharing.

In response to these problems a range of organizations is developing and improving telecommunications infrastructure in Africa (e.g., the African Information Society, the African Development Forum, the African Telecommunications Union, the African Connection, USAID's Leland Initiative, and NOAA's RANET project).

Recommendation: The U.S. government (e.g., USAID and NOAA) should continue to assist African countries in improving telecommunications infrastructure so that large computer files containing geographic data can be readily distributed within national and global spatial data infrastructures.

Collection and Maintenance of Geographic Information and Data

Data Continuity and Technological Uncertainty

For geographic information to be useful for long-term sustainable development and natural resource management, the data source needs to be dependable into the foreseeable future. With the exception of development programs now capitalizing on satellite meteorological observations, most programs will conclude as demonstrations rather than becoming operational, in part because of cost and related uncertainty over future availability of data.

Geographic information technology is rapidly changing. Dramatic changes in architectures, configurations, and approaches to data processing and handling technologies are creating concerns about technological obsolescence. The issues of data continuity and rapid technological change are important considerations when building sustainable geographic information activities. Rapidly evolving technology makes it difficult to provide access to low-cost data analysis tools and to generate continuous datasets. Without some way to assure data continuity (NRC, 1995), investments by development organizations in training and capacity building will be less useful than they could be. Without assurances that these investments will be useful in the future, it will be more difficult for African governments to invest in their own capacity and infrastructure. Changes in data access policy, data cost, or the elimination of an observation program create uncertainties about long-term benefits of international programs to Africans.

Global Positioning Systems

Conclusion: Global Positioning System (GPS) information is broadcast worldwide to virtually anyone in any country, and is of great importance to the practical collection and use of fundamental geographic data for Agenda 21-related initiatives.

Recommendation: The utility of GPS information should not be reduced by reintroducing selective availability, and its continuity should be guaranteed. The U.S. Department of Defense should continue to allow free access to Global Positioning System data.

National Polar-orbiting Operational Environmental Satellite System, Terra, and Landsat

There are low-cost sources of coarse and medium spatial resolution land-cover information for Africa. These come from sensors that include AVHRR (1×1 km), the Moderate Resolution Imaging Spectroradiometer (MODIS) (250×250 m to 1×1 km), and Landsat satellite sensors (79×79 m to 15×15 m). The resultant datasets include Global Land Cover (AVHRR), Tropical Ecosystem Environment Observations by Satellite (AVHRR), GeoCover Land Cover (Landsat), and Africover (Landsat). In addition to global land-cover mapping applications, NOAA's AVHRR is a widely used source of satellite data for cloud and sea-surface temperature mapping in meteorological applications, natural resource management and early warning systems (e.g., FEWS NET, LEWS). This class of sensor flies onboard NOAA's "operational" satellites, and its successor will likely continue operating until 2018 onboard the National Polar-orbiting Operational Environmental Satellite System (NPOESS). NASA's advanced MODIS sensor on its Terra satellite platform has a range of mapping applications including land cover, fire, and productivity.

Conclusion: Land-cover datasets and vegetation indexes are valuable resources for natural resource management and development planning in rural areas. Similar datasets and indexes can be constructed in the future for change detection and many other applications as long as there is continued flow of data from AVHRR, MODIS, and Landsat (or their equivalents).

Recommendation: Until at least 2018, NASA, NOAA, and the U.S. Department of Defense (DOD) should carry out their plan for the National Polar-orbiting Operational Environmental Satellite System to ensure that it supplies relatively coarse spatial and high temporal frequency observations (such as the AVHRR follow-on) that are necessary for the multitude of applications in Africa and elsewhere.

Recommendation: NASA and the U.S. Geological Survey (USGS) should take measures to ensure that the Landsat data continuity mission(s) provides long-term continuous data, perhaps through making the Landsat program an operational system for land observations to support sustainable development and natural resource management in Africa and elsewhere. NASA should also ensure that sensors on its Terra and Aqua satellites (e.g., MODIS, ASTER, AMSR-E) continue to provide data for meteorological and land observation applications.

These actions would address data continuity at the data source. One means of reducing uncertainty in the data at the downstream end is to develop databases using data from more than one source. Such multi-sensor approaches as the twinning of Landsat and Système Pour l'Observation de la Terre (SPOT) satellite imagery for high-spatial-resolution land-cover change information promote flexibility for agencies that base new programs on the availability of data. This approach would benefit from close coordination and cooperation among international data providers and between data providers and donor agencies. Flexibility in the provision of hardware and software technologies would also be necessary; often, programs are tightly coupled to specific, sometimes unique data processing and analysis systems. More use of open, interoperable software environments, as promoted in SDIs, would enhance the flexibility and reduce the vulnerability of these programs.

Focused Geographic Data Requirements

Human Population Distribution

Conclusion: National census data provide the foundation for measuring population distribution and change at the national to local scales. The strengths of human population censuses arise from their completeness of coverage; continuity of statistics from census to census; and the detail that each census provides about population sub-groups in local areas. In the current worldwide development arena, such key issues as good governance, poverty eradication strategies, and the need to promote economic growth with social equity all require population and other demographic data at the detailed local scale that only a population census can provide. Moreover, there exists an increasing demand for disaggregated data at the sub-national level.

Despite these needs, datasets on population distribution from many African censuses are incomplete often owing to high costs and insufficient funds. Progress toward Agenda 21 goals is impeded by this lack of complete, reliable data on human population distribution.

Recommendation: USAID and the U.S. Bureau of the Census should provide financial and technical support to national census offices and bureaus in Africa to help them complete censuses, geographically reference the data, and make the data available in disaggregated form to decision-makers.

Very High Spatial Resolution Remotely Sensed Data

Conclusion: Many Agenda 21 issues concentrate on urban and suburban areas (Chapter 2). Addressing sustainability issues relating to urban and suburban land use (including ownership) and infrastructure requires very high spatial resolution ($\leq 1 \times 1$ m resolution) remotely-sensed data. High-resolution data are costly whether obtained from satellites or airborne sensors. Although there are inexpensive options for obtaining high-quality coarse (1×1 km) and medium (30×30 m) spatial resolution land-cover datasets for parts of Africa, there is no economical method of obtaining very high spatial resolution imagery to inventory and monitor change in urban areas in Africa. "Image grants" would help to inventory and map the continually changing characteristics of urban infrastructures.

Recommendation: USAID should consider purchasing very high spatial resolution images (i.e., $< 1 \times 1$ m) on a regular basis (at 5 to 10 year intervals) for urban areas in Africa and donating them to African organizations to ensure continuity of the data source and change detection. The imagery might include airborne analog or digital photography or satellite-derived high-resolution imagery. The areas surveyed could be requested by African organizations on the basis of importance of the problem and technical and organizational capacity to use the data. One model for this concept is the U.S. Science Data Buy (Box 5-3).

Elevation Data

Elevation (topographic) data have many uses (Table 5-2) but are often inaccurate, of limited extent, or nonexistent, owing to inaccessibility of Earth's mountain ranges, deserts, and forests. Even where access is practical, traditional surveying methods are expensive. Furthermore, neighboring countries may use differing data-collection methods that cause data discontinuities at borders, whereas natural resources (e.g., rivers) often cross these borders. To address these challenges the United States has joined with a number of countries and organizations to produce two digital elevation datasets: the GTOPO30 dataset (with its derivative hydrologic product: HYDRO1K), and the 2000 Shuttle Radar Topography Mission (SRTM) dataset. The GTOPO30 dataset is a global digital elevation dataset whereas the

SRTM dataset covers 80 percent of the globe. In the current plan, which is not finalized, SRTM data will be released at 30 × 30 m spatial resolution for the United States and at 90 × 90 m spatial resolution for the rest of the world.

Conclusion: The GTOPO30 dataset is of limited value in Africa and most other developing countries for monitoring ecosystems, urban and rural infrastructures, and hydrology because of its coarse spatial resolution (1.1 × 1.1 km). Fortunately, elevation data derived from NASA's Shuttle Radar Topography Mission in 2000 may be more suitable for many applications because all data (a) were collected during a single 11-day mission using standardized technology, (b) will have accurate geodetic control, and (c) will be homogeneous across each continent (Chapter 5).

Recommendation: NASA should produce digital elevation data from the Shuttle Radar Topography Mission at the highest possible spatial resolution (e.g., 30 × 30 m) for all areas. The data should be made available without restriction and at affordable cost. NASA should also provide the synthetic aperture radar ortho-image mosaics at 30 × 30 m spatial resolution that are being produced as part of the processing. These mosaics would provide additional information about land-cover conditions and surface roughness characteristics especially in tropical regions shrouded by cloud cover.

Conclusion: A valuable hydrologic product for application to Agenda 21 issues could be derived from the Shuttle Radar Topography Mission with almost global 90 × 90 m (perhaps 30 × 30 m) spatial resolution. This derivative product would have applications at the sub-regional level where the low-resolution (1.1 × 1.1 km) HYDROIK dataset currently is inapplicable.

Recommendation: Serious consideration should be given by the USGS to modeling the Shuttle Radar Topography Mission-derived 30 × 30 m digital elevation data to produce the most accurate, affordable hydrologic network database with global coverage.

Legacy Data

The earliest baseline against which future change can be compared often comes from historical legacy data (e.g., paper maps, and information in monographs, other documents, and verbal histories). In many instances legacy data can be digitized, placed in a GIS, and analyzed in conjunction with more recent geographic data, such as satellite remotely-sensed data. The time scales over which change can be detected are extended through use of legacy data. Additionally, they contain place names and provide valuable insights on ethnicity and population growth. Bridges between local knowledge and modern technology are built through the use of legacy data.

Despite their obvious benefits, legacy data are being lost or remain inaccessible and unused. Efforts are underway to preserve legacy data and ensure that they are used. International organizations including the French Institute of Scientific and Technological Research for Cooperative Development and DEVECOL,[14] and African regional organizations such as the Fundamental Institute of Black Africa in Dakar, Senegal, and the University of Ibadan in Nigeria are working to preserve legacy data and make it available to decision-makers.

***Recommendation*:** To complement these efforts to preserve and enhance the use of valuable legacy data, U.S. government agencies (e.g., USAID and USGS) should assist African countries and organizations to identify, integrate, and maintain existing sources of information (legacy data). They should also provide African countries with copies of such legacy data as reports, maps, statistics, aerial and satellite photographs, and other relevant data and materials currently held outside those countries. The first task would be substantial, whereas the second would be more routine once the first is addressed.

Cadastral Data

Owning land provides individuals with economic assets that can be traded in land markets, used as collateral to raise credit or as security for various forms of economic improvements. Because individual land ownership is nonexistent in large parts of rural Africa, except in eastern and southern Africa, challenges remain for rural Africans to obtain credit from lending institutions in their bid to improve quality of life.

Conclusion: The production of cadastres[15] is costly and has been a low priority for most African countries and donor agencies, even when there are clear benefits. GPS, used in concert with GIS, produces cadastral data more cheaply than traditional surveying techniques and will facilitate production of cadastres. Continued, cautious development of cadastres will facilitate land management and administration, promote greater efficiency in the operation of land markets, strengthen the operations of free-market economies, and reinforce the ability of governments to initiate and sustain land and agrarian reforms (e.g., de Soto, 2000).

Over time, cadastres could play an indirect role in poverty reduction, especially through enhancing access to credit facilities and providing socioeconomic information for effective settlement management. Many Africans have no easily located residential addresses to facilitate their effective

[14]A contraction of DEVelopment ECOLogy: an information resource organization in the United States.

[15]A cadastre is a map accompanied by a register showing the ownership or possession of individual units of land.

participation in social and economic transactions (ECA, 2001). These inadequacies have been one reason why the systematic delivery, management, expansion, and improvement of services to all segments of the population, the effective collection of taxes and rates, and the cost recovery for utilities and services have been difficult to implement in urban areas (ECA, 2001).

Recommendation: Because of the potential of cadastres to address Agenda 21 issues including poverty reduction and land resource management, the U.S. government (USAID and USGS) should assist African countries to develop cadastres.

Decision-Support Systems

Conclusion: Decision-support in the area of land cover (Chapter 6) will be one of the most fruitful application areas of geographic data and tools in Africa. The livelihoods of the majority of Africans depend on agriculture and natural resources, and there are many pressing problems within these sectors. Addressing these problems demands better data and better ways of analyzing the relationship between human activities and changes on the land surface. International activities could accelerate the use of decision-support systems for land-cover applications in Africa. Strategies to improve or create these datasets are needed, and these strategies would work best when built on existing initiatives and networks.

Recommendation: An effective land-cover decision-support system should include a standard classification system; baseline data and change detection capabilities; hot spot detection and high risk zone prediction capabilities; analysis and modeling of proximate (mainly human) causes of change; linkages between direct observations, case studies, and models; and established environmental indicators.

Geospatial Capacity Building for Sustainable Development

Coordination and Partnering for Meeting African Data Needs

Conclusion: Moving beyond the current state of the art in the application of geographic data in Africa will require greater coordination among data providers, donor agencies, and the science community and end-users in Africa. Already the requirements for the next generation of remote-sensing systems are being defined or developed, yet there appears to be little dialog between the space agencies and the donor agencies, and even less input from potential end-users of the data in Africa. Lessons learned in the application of existing data for decision-making could be fed back into the definition of future observation and data system requirements, particularly in government science agencies.

Recommendation: Data providers, U.S. government agencies, and partners should work closely with African organizations to define and integrate the data needs of Africans into future programs (e.g., for new satellite remote-sensing missions), and to maximize efficiency of new programs through a coordinated approach.

Partnerships for Capacity Building

Conclusion: Partnerships promote sharing of resources, improved communication and cooperation, and acceptance of shared standards required for spatial data infrastructures. Effective use of geographic information science in sustainable development will be associated with the strengthening of existing partnerships and the emergence of new forms of partnerships involving universities, industry, government, and civil society. Partnerships among universities and the private sector in geospatial capacity building are key to achieving a balance between supply and demand for geographic information, tools, and services in Africa. Research networks that develop as a result of these partnerships promote broad exchange of information on sustainable development including best practices. Development of effective partnerships requires the support and incentives of both African and international donor governments.

Recommendation: In promoting organizational cooperation, emphasis should be placed on fostering innovation and the transfer of geographic data and technology through: (1) partnerships and research networks among government agencies, research and training institutions, the private sector, and the non-governmental sector; (2) international collaboration involving developed and African countries; and (3) cooperation between African and other developing countries.

Human and Organizational Capacity

Most of the existing geographic information activities in Africa were initiated in response to humanitarian needs such as famine and natural disaster, and implemented through focused, time-limited projects. Learning to apply modern geographic information and tools to address evolving societal needs requires a long learning period and attention to the development of societal capacity in science and technology. Universities are the logical source of this kind of education and training because they focus on teaching and research. With the appropriate policies and incentives, universities are also natural incubators for enterprises and social organizations. The organizations of civil society are important in many African countries where the functions of the state are inadequate.

Conclusion: It is unlikely that long-term capacity building in technical fields such as geographic information science can be sustained in the absence of strong foundations in higher education with emphasis on science and technology. Despite the difficulties African universities face, they remain vital to the generation of new knowledge and have the potential for organizational capacity building. The application of geographic information to sustainable development will depend on the quality, character, and direction of university education in Africa. There is an urgent need to coordinate and strengthen the capacity of university departments providing both research and training in geographic information science.

Recommendation: African universities should become a focus for capacity-building including training and research in geographic information science, and development organizations should coordinate their efforts to achieve this goal.

Conclusion: A cadre of well-trained individuals will need to be formed in each country to apply geographic data and information in support of sustainable development in Africa.

Recommendation: Continuing and on-the-job training should become an integral part of the process of enhancing geospatial capacity. Organizations that provide professional training in geographic information sciences, such as regional centers and polytechnics, should be strengthened.

SUMMARY

Geographic data lie at the heart of many Agenda 21 issues. These data are already in use in a growing geographic information community in Africa. Although there exist a number of barriers to effective application of geographic data to Agenda 21 issues, it is likely that demand for these data will quicken the pace toward the disappearance of these barriers.

REFERENCES

Burrough, P. A. 1986. Principles of Geographic Information Systems for Land Resource Assessment. Monographs on Soil and Resources Survey No. 12. New York: Oxford University Press.

Colwell, R. N. 1997. History and place of photographic interpretation. Pp 33–48 in Manual of Photographic Interpretation, W. R. Philipson, ed. Bethesda, MD: American Society for Photogrammetry and Remote Sensing.

Cowen, D. J. 1988. GIS versus CAD versus DBMS: What are the differences? Photogrammetric Engineering and Remote Sensing 54:1551-1554.

De Soto, H. 2000. The Mystery of Capital: Why Capitalism Triumphs in the West and Fails Everywhere Else. New York: Basic Books.

Dernbach, J. C. 2002. Sustainable Development: Now More Than Ever. Environmental Law Reporter 32 (1):10003-10019.

ECA (UN Economic Commission for Africa). 2001. The Future Orientation of Geoinformation Activities in Africa: A Position Paper. Available at <http://www.uneca.org/eca_resources/Conference_Reports_and_Other_Documents/disd/codi/docs/doc3EN.pdf>. Accessed May 15, 2002.

EIS-Africa (Environmental Information Systems-Africa). 2001. Environmental Information Systems Development in Sub-Saharan Africa: Approaches, Lessons and Challenges. E. Gavin and J. Gyamfi-Aidoo, eds. EIS-Africa.

Fukuda-Parr, S., C. Lopes, and K. Malik. 2002. Capacity for Development: New Solutions to Old Problems. London: Earthscan Publications.

Geography Education Standards Project. 1994. Geography for Life: National Geography Standards. Washington, DC: National Geographic Society.

GISD (Geographic Information for Sustainable Development). 2002. Geographic Information for Sustainable Development. Available at <http://www.opengis.org/gisd>. Accessed May 15, 2002.

Jensen, J. R. 2000. Remote Sensing of the Environment: An Earth Resource Perspective. Saddle River, NJ: Prentice-Hall.

NRC (National Research Council). 1995. Earth Observations from Space: History, Promise, and Reality. Washington, DC: National Academy Press.

NRC. 1999. Our Common Journey: A Transition Toward Sustainability. Washington, .C: National Academy Press.

NRC. 2002. Community and Quality of Life: Data Needs for Informed Decision Making. Washington, DC: National Academy Press.

Star, J. L. and J. Estes. 1990. Geographic Information Systems: An Introduction. Englewood Cliffs, NJ: Prentice-Hall.

UN (United Nations). 1997. Earth Summit: Rio + 5. Country Profiles–African States. Available at <http://www.un.org/esa/earthsummit/afri-cp.htm>. Accessed May 17, 2002.

UN. 2001. SADC Progress Report on the Implementation of Agenda 21 and Sustainable Development: A Report to the 2002 World Summit on Sustainable Development. Available at <http://www.johannesburgsummit.org/web_pages/sadc_prepcom_progress_report.pdf>. Accessed August 1, 2002.

UNCED (United Nations Conference on Environment and Development). 1992. Report of the United Nations Conference on Environment and Development, Rio de Janeiro, 3-14 June 1992. Rio de Janeiro: UN. (Annex I: Rio Declaration; Annex II: Agenda 21).

UNDP (United Nations Development Program). 1997. Governance for Sustainable Development. Policy paper. Available at <http://magnet.undp.org/policy.default.htm>. Accessed May 20, 2002.

WCED (World Commission on Environment and Development). 1987. Our Common Future. New York: Oxford University Press.

World Bank. 2002. Data and Statistics. Available at <http://www.worldbank.org/data/countrydata/countrydata.html>. Accessed June 5, 2002.

1

Introduction

In 1992, world leaders adopted Agenda 21 (UNCED, 1992), the work program of the 1992 U.N. Conference on Environment and Development in Rio de Janeiro. The landmark event provided a political foundation and action items (Table 1-1) to facilitate the global transition toward sustainable development (Box 1-1). The international community is marking the tenth anniversary of this conference by holding the World Summit on Sustainable Development in Johannesburg, South Africa, in August 2002. The main goals of the summit are to "reinvigorate the global commitments to, and achieve a higher level of international solidarity and partnership in the promotion of sustainable development (UN, 2001)." The summit builds on the political momentum created by the U.N. Millennium Declaration, in which world leaders committed themselves to achieving a broad range of time-bound international sustainable development objectives for which sustainable development provides a unifying framework (UN, 2001).

This report is a component of the U.S. State Department's contribution to the World Summit on Sustainable Development, the "Geographic Information for Sustainable Development" project (GISD, 2002). Because South Africa is hosting the summit, it seemed appropriate for this report to focus on sub-Saharan Africa. Examples are drawn from case-study regions where the U.S. Agency for International Development and other agencies have broad experience. Although African countries are the geographic focus of the report (Figure 1-1), the material in the report has broader applicability. This chapter describes the purpose of the report, presents the committee's charge, provides background on major themes, and outlines the organizational structure.

PURPOSE OF THIS REPORT

This report summarizes the importance and applicability of geographic[1] data for sustainable development. Geographic data describe spatial variations across the landscape at a variety of scales (local, national, global) and include such elements as climate, elevation, soil, vegetation, population, land use, and economic activity. The report draws on experiences in African countries and examines how future sources and applications of geographic data could provide reliable support to decision-makers[2] as they work towards sustainable development. The committee emphasizes the potential of new technologies, such as satellite remote-sensing systems[3] and geographic information systems (GIS), that have revolutionized data collection and analysis, over the last decade.

The Charge to the Committee

As a component of the State Department's contribution to the World Summit in Johannesburg (August 2002), this study will examine the geographic foundation for natural resource management and development issues in Africa. Centered on a place-based,[4] integrative framework, the study draws on experiences of U.S. government agencies, international

[1]*Geography* is an integrative discipline that brings together the physical and human dimensions of the world in the study of people, places, and environments. Its subject matter is the Earth's surface and the processes that shape it, the relationships between people and environments, and the connections between people and places (Geography Education Standards Project, 1994).

[2]*Decision-makers* in the context of sustainable development choose actions that directly or indirectly affect the environment and reside in all levels of government, the citizenry, the private sector, non-governmental organizations, or overseas development agencies, for example.

[3]*Remote-sensing* is "the measurement or acquisition of information of some property of an object or phenomenon by a recording device that is not in physical or intimate contact with the object or phenomenon under study" (Colwell, 1997). Satellites and aircraft are common platforms for remote-sensing systems.

[4]*Place-based* studies are "the systematic analysis of social, economic, political, and environmental processes operating in a place that provides an integrated understanding of its distinctiveness or character" (NRC, 1999).

TABLE 1-1 Action Items from Agenda 21[a]

SECTION I. SOCIAL AND ECONOMIC DIMENSIONS

2. International cooperation to accelerate sustainable development in developing countries and related domestic policies
3. Combating poverty
4. Changing consumption patterns
5. Demographic dynamics and sustainability
6. Protecting and promoting human health conditions
7. Promoting sustainable human settlement development
8. Integrating environment and development in decision-making

SECTION II. CONSERVATION AND MANAGEMENT OF RESOURCES FOR DEVELOPMENT

9. Protection of the atmosphere
10. Integrated approach to the planning and management of land resources
11. Combating deforestation
12. Managing fragile ecosystems: combating desertification and drought
13. Managing fragile ecosystems: sustainable mountain development
14. Promoting sustainable agriculture and rural development
15. Conservation of biological diversity
16. Environmentally sound management of biotechnology
17. Protection of the oceans, all kinds of seas, including enclosed and semi-enclosed seas, and coastal areas and the protection, rational use and development of their living resources
18. Protection of the quality and supply of freshwater resources: application of integrated approaches to the development, management and use of water resources
19. Environmentally sound management of toxic chemicals, including prevention of illegal international traffic in toxic and dangerous products
20. Environmentally sound management of hazardous wastes
21. Environmentally sound management of solid wastes and sewage-related issues
22. Safe and environmentally sound management of radioactive wastes

SECTION III. STRENGTHENING THE ROLE OF MAJOR GROUPS

24. Global action for women toward sustainable and equitable development
25. Children and youth in sustainable development
26. Recognizing and strengthening the role of indigenous people and their communities
27. Strengthening the role of non-governmental organizations: partners for sustainable development
28. Local authorities' initiatives in support of Agenda 21
29. Strengthening the role of workers and their trade unions
30. Strengthening the role of business and industry
31. Scientific and technological community
32. Strengthening the role of farmers

SECTION IV. MEANS OF IMPLEMENTATION

33. Financial resources and mechanisms
34. Transfer of environmentally sound technology, cooperation and capacity building
35. Science for sustainable development
36. Promoting education, public awareness and training
37. National mechanisms and international cooperation for capacity building in developing countries
38. International institutional arrangements
39. International legal instruments and mechanisms
40. Information for decision-making

[a]The numbers indicate the chapter of Agenda 21 that relates to each action item. Chapters 1 and 23 are preambles, and therefore are not included in this table.

SOURCE: UNCED (1992).

BOX 1-1
Sustainable Development: A Matter of Definition

The concept of "sustainable development" has a long history in scientific thought. As early as 1749 the Swedish botanist Linnaeus in his *Oeconomia naturae*, linked economy to nature in a way that resembles many of the concepts of sustainable development. His economic program focused on the need to make efficient use of existing resources rather than pursue military expeditions as a means of economic survival. Over 200 years later, in *Our Common Future*, the World Commission on Environment and Development (WCED, 1987) gave international prominence to sustainable development and defined it as "development that meets the needs of the present without compromising the ability of future generations to meet their own needs." *Our Common Journey*, a report of the U.S. National Research Council (NRC, 1999), added a temporal dimension, defining sustainable development as "the reconciliation of society's developmental goals with its environmental limits over the long term." These definitions reflect the growing need to provide an ethical framework for integrating developmental and environmental goals. Sustainable development is therefore a set of guiding principles whose implementation is reflected in a variety of action programs, of which Agenda 21 is the most prominent.

FIGURE 1-1 Countries of Africa. Courtesy D. Zimble, ESRI.

groups, decision-makers, and experts to examine (1) existing remote-sensing and GIS efforts in case-study regions, and lessons learned from those efforts; (2) existing levels of local expertise and technology, and ongoing efforts in geospatial capacity building[5]; (3) a range of questions relating to the practical application of new and existing spatial data (e.g., required resolution, challenges of integrating layers of environmental and social data, and baseline data against which future change can be measured); (4) the role of decision-support systems in the application of these data; and (5) options for making efforts sustainable beyond 2002.

The geographic foundation for Agenda 21 in Africa involves a wide array of geographic data, tools, and perspectives (including social, environmental, and economic data; maps and models; and the analysis of pattern and processes, place and scale). This study was undertaken in support of the U.S. Geographic Information for Sustainable Development alliance that focuses on the uses of earth observation data and GIS to address Agenda 21 issues. In this context the report emphasizes fundamental data types that are needed in many applications and ways of increasing accessibility to these data in Africa. An assessment of the potential applications of the full range of geographic data, tools, and concepts for natural resource management and development in Africa would also be valuable.

GEOGRAPHIC DATA AND SUSTAINABLE DEVELOPMENT

Chapter 40 in Agenda 21, "Information[6] for Decision-Making," stresses the need for more and different types of data to be collected at all scales to track the status and trends of Earth's ecosystems, natural resources, pollution, and socioeconomic variables. Chapter 40 concludes that "the gap in the availability, quality, coherence, standardization and accessibility of data between the developed[7] and the developing world has been increasing, seriously impairing the capacities of countries to make informed decisions concerning environment and development."[8]

Five years after UNCED in a 1997 assessment of Agenda 21 action items (UN, 1997), most of the nine responding African governments[9] described their existing databases as "poor" or containing "some good data but many gaps." Ten years after UNCED the work of the U.N. Economic Commission for Africa, which presses for greater awareness of the significance of geographic information in socioeconomic development among African governments and other sectors[10] (ECA, 2001), illustrated that much remained to be achieved in applying geographic information to sustainable development.

Geographic data are obtained from ground-based (in situ) measurements or from remote-sensing systems. These data are of little practical value in sustainable development decision-making if they cannot be analyzed in conjunction with development data, such as economic or health data, that are geographically referenced[11] (Jensen, 2000). Data that describe environment and development can be linked by geographic location to provide greater understanding of complex issues, and GISs were developed specifically for this purpose. A GIS is formed from a set of map layers or overlays registered to a common geographic coordinate system (Figure 1-2) and is "a digital information system that is designed to work with data referenced by spatial or geographic coordinates" (Star and Estes, 1990). Geographic information systems are powerful tools for ingesting, storing, retrieving, transforming, processing, and displaying geographic data (Burrough, 1986). If a GIS involves the integration of geographically referenced data in a problem-solving situation, it can become important for decision-making, or a "decision-support system" (Cowen, 1988). Geospatial capacity is essential for all these steps so that full use can be made of the capabilities of geographic data for supporting sustainable-development decision-making.

Geographic data provide information for sustainable development across many sectors including agriculture and aquaculture, industry, mining, health, transportation, trade, and tourism (Table 1-2). These data can contribute to implementing Agenda 21 in Africa. Among the Agenda 21 issues (Table 1-1), many African nations, international organizations, and African non-governmental organizations have identified poverty and the unequal distribution of benefits of development as the most important in Africa today. Poverty eradication will require equitable economic and social development within sustainable environmental parameters. The

[5]Capacity is the ability to undertake certain activities, solve problems, and achieve objectives (Fukuda-Parr et al., 2002), such as interpreting a map (an example of human capacity), possessing a computer (organizational capacity), or sharing data (societal capacity). Capacity is built by enhancing these abilities.

[6]*Information* is data that humans assimilate and evaluate to solve a problem or make a decision (EIS-Africa, 2001).

[7]Developed countries are those with "high" gross national income (World Bank, 2002). Developing countries are those with "low" or "medium" gross national income. The committee adopted these terms in the report.

[8]*Development* includes at least four related concepts (Dernbach, 2002): peace and security, economic development, social development, and national governance that secures peace and development.

[9]Algeria, Benin, Egypt, Gabon, Malawi, South Africa, Tunisia, Uganda, and Zimbabwe.

[10]The committee applies the term "sector" to groupings such as environmental, social, and economic sectors.

[11]These are data with known latitude, longitude, and elevation, or other horizontal and vertical coordinates.

FIGURE 1-2 A geographic information system consists of a number of geographic data layers that are linked to one another in a common geographic coordinate system (such as latitude and longitude). In this example the various thematic layers consist of remotely sensed data (from Landsat Thematic Mapper), the hydrologic network, digital elevation, the road network, and watershed boundaries.

environment is the fundamental matrix providing natural resources, waste assimilation, and links between people and the natural world (NRC, 2002).

STRUCTURE OF THE REPORT

The report is designed as a resource to develop an appreciation for the applicability of geographic data to sustainable development issues and to highlight a path toward integration of geographic information into the decision-making process. Its structure is in three parts.

- Introductory material (Chapters 2-3);
- Technology: Infrastructure, data sources, and tools (Chapters 4-7); and
- Capacity building (Chapter 8).

Chapter 2 describes progress towards sustainable development in a global and African context, and the main challenges for implementation of Agenda 21. Chapter 2 ends with a discussion of the value of geographic information for implementing Agenda 21. Chapter 3 introduces the case-study regions in sub-Saharan Africa and describes ongoing activities in these regions that involve remote-sensing and GIS.

Chapter 4 examines the status and prospects of infrastructures that facilitate broad use of geographic information. Spatial data and telecommunications are the infrastructures described. Chapters 5 and 6 summarize fundamental and supplemental sources of geographic data and their adequacy for Agenda 21 applications. These chapters are supported by technical annexes, and by a list of acronyms in Appendix D. Chapter 7 discusses decision-support tools (in particular GIS) that assist users in converting geographic data into valuable information for decision-makers.

The final section (Chapter 8) examines geospatial capacity building—human, organizational, and societal. It looks at ongoing efforts and ways geospatial capacity building may ensure that countries in Africa continue on a path toward sustained use of geographic information in Agenda 21 action items.

Specific examples from the four case-study regions and applications of geographic data and tools to sustainable development challenges in Africa are discussed throughout the report. Chapter 3 presents in tables a broad range of applications in the agriculture and natural resource sectors, environmental monitoring, and demography. Human health and food security are prominent issues in Africa, and Chapters 6 and 7 discuss data for managing contagious and vector-borne disease (e.g., malaria), famine early warning, and early warning systems in forestry and livestock management. Chapter 7 also describes a decision-support system for wildlife management. The committee's recommendations appear in Chapters 4 through 8, and are summarized in a concluding chapter (Chapter 9).

TABLE 1-2 Common Issues for Development Planners and Natural Resource Managers in Africa and Applications of Geographic Data and Information

Issue	Geographic Data and Information Needs
Land classification and land-use planning	• Elevation data • Vegetation cover • Soil data • Climatic data
Land allocation (allocate land to citizens for various uses)	• Land classification information to ensure compatibility of use with allocated land • Population statistics
Land resource use and management (ensuring land is used for designated use; planning for future use)	• Cadastral data showing allocated users and uses • Administrative land records • Legal land registers
Infrastructure and urban management (efficient and equitable provision of urban services)	• Population statistics • Location of existing services: waste disposal, water, and power installations (and statistics thereon) • Urban road networks
Transportation (providing energy-efficient and safe transportation systems)	• Elevation data • Geophysical data • Statistics of trips between population and employment centers
Tourism (developing and promoting eco-tourism)	• Location of existing tourist facilities • Statistics on tourist preferences and capacities of facilities • Decision support for land-use decisions
Sewage discharge (prevention of discharge of untreated sewage into water bodies)	• Location and capacities of existing sewage treatment plants • Statistics on population centers
Water resource management and conservation (to ensure sustained use of water for domestic, industrial, and agricultural uses)	• Location of water bodies and courses with flow and condition data • Location of water users • Meteorological data on precipitation, evaporation • Geophysical data on rock formations • Water consumption statistics
Pollution prevention/management	• Location of potential sources of pollution • Location of pathways of pollutant dispersal (e.g., water courses) • Data on products of industry and input raw materials
Land degradation (to enforce land-use practices, ensure security of tenure, combat desertification)	• Land-use data • Distribution of land cover • Land allocation data • Land tenure data • Population data • Climate data • Soils data • Data on management practices
Rangeland and livestock management	• Distribution of agro-ecological zones • Livestock statistics
Forest resource management	• Land-use data and classifications • Vegetation cover • Census of endangered species • Population data • Location of fuel-energy-intensive industries
Biological diversity (conservation of unique and endangered species; management of national parks, nature reserves, protected areas)	• Distribution and concentration of unique flora and fauna • Illegal traffic in endangered species • Land cover change • Elevation distribution • Transportation infrastructure
Energy resource management (to reduce dependence on biomass energy, and to develop renewable energy sources)	• Population and socio-economic data, including energy demand • Distribution of forest resources • Climatic data for solar and wind power developments

continues

TABLE 1-2 Continued

Issue	Geographic Data and Information Needs
Mineral resource management	• Distribution of mineral resources • Administrative data on exploration and prospecting activities
Human health management (equitable distribution of health facilities; early diagnosis, prompt treatment and environmental management; hygiene education)	• Demographic data • Location and capacity of existing facilities and population centers • Distribution of disease vectors • Distribution of water pockets for remedial action • Distribution, flow, and chemical characteristics of water courses

SOURCE: Adapted from ECA (2001)

REFERENCES

Burrough, P. A. 1986. Principles of Geographic Information Systems for Land Resource Assessment. Monographs on Soil and Resources Survey No. 12, New York: Oxford University Press.

Colwell, R. N. 1997. History and place of photographic interpretation. Pp 33–48 in Manual of Photographic Interpretation, W. R. Philipson, ed. Bethesda, MD.: American Society for Photogrammetry and Remote Sensing.

Cowen, D. J. 1988. GIS versus CAD versus DBMS: What are the differences? Photogrammetric Engineering and Remote Sensing 54:1551-1554.

Dernbach, J. C. 2002. Sustainable Development: Now More Than Ever. Environmental Law Reporter 32 (1):10003-10019.

ECA (UN Economic Commission for Africa). 2001. The Future Orientation of Geoinformation Activities in Africa: A Position Paper. Available at <http://www.uneca.org/eca_resources/Conference_Reports_and_Other_Documents/disd/codi/docs/doc3EN.pdf>. Accessed May 15, 2002.

EIS-Africa (Environmental Information Systems-Africa). 2001. Environmental Information Systems Development in Sub-Saharan Africa: Approaches, Lessons and Challenges, E. Gavin and J. Gyamfi-Aidoo, eds. EIS-Africa.

Fukuda-Parr, S., C. Lopes, and K. Malik. 2002. Capacity for Development: New Solutions to Old Problems. London: Earthscan Publications.

Geography Education Standards Project. 1994. Geography for Life: National Geography Standards. Washington, DC: National Geographic Society.

GISD (Geographic Information for Sustainable Development). 2002. Geographic Information for Sustainable Development. Available at <www.opengis.org/gisd>. Accessed May 15, 2002.

Jensen, J. R. 2000. Remote Sensing of the Environment: An Earth Resource Perspective. Saddle River, NJ: Prentice-Hall.

NRC (National Research Council). 1999. Our Common Journey: a Transition Toward Sustainability. Washington, DC: National Academy Press.

NRC. 2002. Community and Quality of Life: Data Needs for Informed Decision Making. Washington, DC: National Academy Press.

Star, J. L. and J. Estes. 1990. Geographic Information Systems: An Introduction. Englewood Cliffs, NJ: Prentice-Hall.

UN (United Nations). 1997. Earth Summit: Rio + 5. Country Profiles–African States. Available at <http://www.un.org/esa/earthsummit/africp.htm>. Accessed May 17, 2002.

UN. 2001. SADC Progress Report on the Implementation of Agenda 21 and Sustainable Development: A Report to the 2002 World Summit on Sustainable Development. Available at <www.johannesburgsummit.org/web_pages/sadc_prepcom_progress_report.pdf>.

UNCED (United Nations Conference on Environment and Development). 1992. Report of the United Nations Conference on Environment and Development, Rio de Janeiro, 3-14 June 1992. Rio de Janeiro: UN. (Annex I: Rio Declaration; Annex II: Agenda 21).

WCED (World Commission on Environment and Development). 1987. Our Common Future. New York: Oxford University Press.

World Bank. 2002. Data and Statistics. Available at <http://www.worldbank.org/data/countrydata/countrydata.html>. Accessed June 5, 2002.

2

Agenda 21 Implementation: Progress, Challenges, and the Role of Geographic Data

INTRODUCTION

This chapter provides an overview of the progress that has been made in implementing Agenda 21. Specifically, it examines national and local Agenda 21 activities, first for the world as a whole and then for African countries. The chapter looks broadly at what has been accomplished and discusses challenges to Agenda 21 implementation. Finally, the chapter summarizes the potential contribution of geographic data to achieving Agenda 21 objectives.

GLOBAL PROGRESS WITH IMPLEMENTATION OF AGENDA 21

The 1992 U.N. Conference on Environment and Development (UNCED) provided a political foundation and programs to facilitate the transition toward sustainable development (UNCED, 1992) (Table 1-1). Agenda 21 was accepted by more than 178 governments in 1992. Progress toward achieving Agenda 21 goals is occurring, but has been slower than anticipated (UN, 2002a). Efforts to integrate environment and development into a common sustainable development framework remain in the early stages (UN, 2002a). Meanwhile, environmental degradation and poverty at a global scale is worsening (Dernbach, 2002).

Three major trends characterize progress on Agenda 21 implementation over the last 10 years. First, the concept of sustainable development prompted a shift from focusing on single issues toward appreciating complex interactions between a wide range of environmental and developmental factors. This shift is part of what has been called "the transition to sustainability" (NRC, 1999). Second, there is a movement from international top-down norm-setting to national institution-building and more "grassroots" approaches at the local government level. Third, Agenda 21 demands place-based scientific and technical knowledge, which has resulted in increasing involvement of research-based institutions such as universities and private enterprises (Juma, 2002).

National Government Implementation of Agenda 21

At the present time at least 85 countries have developed national strategies for implementing Agenda 21 (UN, 2002a). Some common strategies include the development of the following:

- thematic policies that articulate broad sustainable development objectives;
- traditional master plans based on national planning cycles;
- mechanisms for coordination with donors; and
- strategies to address international obligations to integrate environmental considerations into thematic activities.

In a number of developing countries sustainable development principles have been introduced in existing national frameworks, such as conservation strategies, environmental plans, national vision statements, and national Agenda 21 initiatives (UN, 2002a).

Local Government Implementation of Agenda 21

The role of local governments is critical in educating, mobilizing, and responding to the public (Lake, 2000). Chapter 28 of Agenda 21 focuses on the local government role. The contents of that chapter provide the framework for the International Council for Local Environment Initiatives (ICLEI) (Box 2-1).

The U.N. undertook two surveys of Local Agenda 21 implementation. The first survey was in 1997, the second in 2001 (UN, 2002b). For the purposes of the surveys "Local Agenda 21," or LA21, was defined as

BOX 2-1
Roles of ICLEI Projects in Local Agenda 21 Implementation

1. *Local Agenda 21 Incentive Grants Project*

Between 1997 and 2000 the Local Agenda 21 (LA21) incentive grants project provided grants, training, and program support to LA21 planning initiatives in 18 cities in Africa, Latin America, and Europe. ICLEI, in partnership with the Open Society Institute, designed the project to learn how open societies can be fostered and how quality of life can be improved.

2. *African Sustainable Cities Network*

The African Sustainable Cities Network (ASCN) aims to build the capacity of local governments to institute participatory environmental planning as an integrated function of public administration. By 2000, 31 African cities in 9 countries (Ghana, Kenya, Malawi, Namibia, South Africa, Tanzania, Uganda, Zambia, and Zimbabwe) were participating in the ASCN and had signed the local government resolutions. The ASCN also focuses on capacity building and exchanges between these cities (and between African and European cities).

3. *European Sustainable Cities and Towns Campaign*

The European sustainable cities and towns campaign unites and assists local governments in engaging in LA21 activities. The campaign aims to promote sustainable development at the local level by strengthening partnerships. The project was launched following the First European Conference on Sustainable Cities and Towns held in Aalborg, Denmark, in May 1994, in the course of which the Aalborg Charter was adopted. To date, over 1,500 local and regional governments from 38 European countries have signed the charter. The campaign is the largest European initiative for LA21.

4. *Local Agenda 21 Charters Project*

Between 1997 and 2000 the Local Agenda 21 charters project aimed to establish partnerships between local governments in developed and developing countries to assist each other in the implementation of their LA21 action plans. Six African countries participated in the project: Ghana, Kenya, Namibia, South Africa, Tanzania, and Zimbabwe. Support was provided through regional training and technical assistance programs and the creation of a global monitoring and reporting system. Between local governments the assistance programs were linked through sustainable development agreements or Local Agenda 21 charters.

SOURCES: <http://www.iclei.org/la21/igp/>, <http://www3.iclei.org/la21/ascn/ascnsum.htm>, <http://www.sustainable-cities.org/sub2.html>; <http://www.iclei.org/europe/suscam.htm>, and <http://www.iclei.org/la21/charters.htm>.

> a participatory, multi-stakeholder process to achieve the goals of Agenda 21 at the local level through the preparation and implementation of a long-term, strategic plan that addresses priority local sustainable development concerns. (UN, 2002b).

By the end of 2001 nearly 6,500 local governments in over 100 countries were involved in LA21 (Table 2-1). Of these local governments 44 percent had active programs and the remainder had committed to the process. In the four years between surveys the number of LA21 activities more than tripled (driven primarily by activities in Europe), and the number of participating countries nearly doubled. LA21 initiatives often have evolved at the local level in the absence of a national campaign. Indeed, 59 percent of the initiatives progressed without national-level impetus.

Progress can also be measured by an increase in the number of LA21 processes that have moved from the vision statement stage into the action planning stage—from 38 percent in 1997 to 61 percent in 2001 (Table 2-2). Most of these have focused on the environment (Figure 2-1). Prominent environmental issues addressed include air quality and water resources management (Table 2-3). Over the next three to five years the prominence of natural resources management issues will increase (Table 2-3).

TABLE 2-1 Number of Local Governments Involved in Local Agenda 21 Activities in December 2001, by Region

Region	Number of Countries	Number of Local Governments
Africa	28	151
Asia-Pacific	17	674
Europe	36	5,292
Latin America	16	114
Middle East	9	73
North America	2	101

SOURCE: UN (2002b).

Private Sector Implementation of Agenda 21

Companies, especially multinational corporations, dominate the transformation of natural resources into products and services. Increasingly they find it in their own interests to meet sustainable development goals. The World Business Council for Sustainable Development maintains a collection of case studies of the sustainability transition from a wide variety of firms (WBCSD, 2002). The private-sector re-

TABLE 2-2 Status of Local Agenda 21 Activities in 2001

LA21 Municipal Planning Documents	Vision Statement	Local Action Plan	Sustainable Development Policy	Monitoring Report
Municipalities completing document	52%	61%	39%	34%
Average year adopted	1999	1999	1998	1999
Documents developed with stakeholder participation	83%	89%	77%	63%
Level of community participation[a]	Medium	Medium	Medium	High
Documents using indicators	52%	55%	51%	70%
Average document time frame[b]	4 years	4 years	2 years	1 year

[a]Based on ranking participation from 1 to 5.

[b]The average number of years for which the vision statement, local action plan, sustainable development policy, or monitoring report is valid.

SOURCE: UN (2002b).

sponse includes technological innovation, implementation of eco-efficiency standards, adoption of social responsibility practices, adjustment of management approaches, and promotion of dialogue and partnerships (Box 2-2).

Global Challenges in Implementing Agenda 21

The goals of Agenda 21 have met with limited success (UN, 2002a). According to the U.N., challenges to the implementation of the Agenda 21 objectives are

- *a fragmented approach toward sustainable development that de-couples environment and development.* The concept of sustainable development integrates environment and development in the long term (Box 1-1). The current fragmented approaches to sustainable development arise from policies and programs at national and international levels that often are short-term and inadequately merge environmental and developmental considerations (UN, 2002a).
- *a lack of integrated national policies and approaches in the areas of finance, trade, investment, technology, and sustainable development.* Commonly, short-term considerations are placed above the long-term use of natural resources, and policies are often compartmentalized.

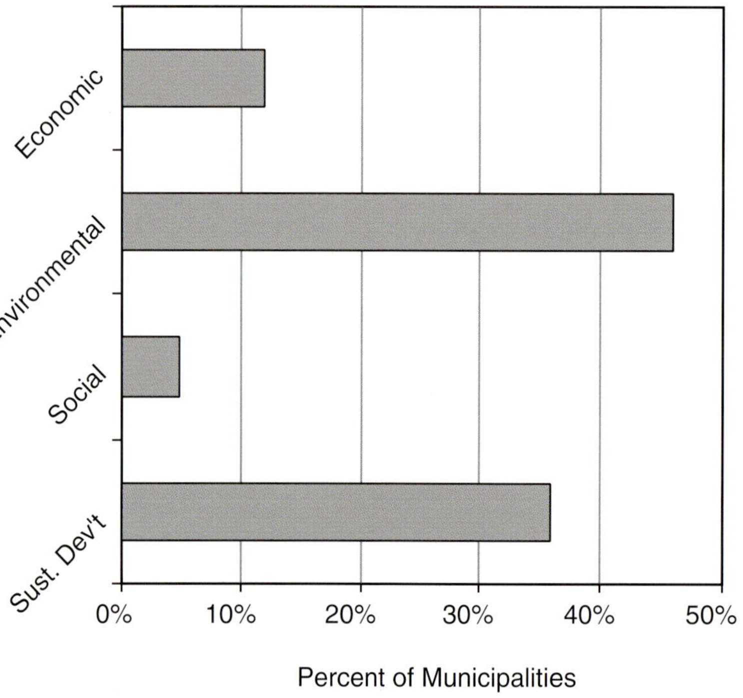

FIGURE 2-1 Focus of local Agenda 21 activities. Sust. Dev't = Sustainable Development. SOURCE: Adapted from UN (2002b).

TABLE 2-3 Local Agenda 21 Activities Underway and Future Priority Issues[a]

Rank	Activities Currently Underway Globally	Priority Issues for the Next 3-5 Years Globally
1	Air quality	Natural resources management
2	Water resources management	Air quality
3	Energy management	Water resources management
4	Transportation	Energy management
5	Natural resources management	Transportation

[a]Respondents were allowed to select more than one option.

SOURCE: UN (2002b).

- *continuing unsustainable patterns of consumption and production that imperil natural life-support systems.* Unsustainable consumption and production is rooted in value systems that drive how natural resources are used. Such value systems have proven slow to adapt.
- *inadequate financial resources and technology transfer from developed countries.* Overseas development assistance has declined over the last decade, and debt burdens limit the opportunities of many developing countries to address sustainability issues. Additionally, private-sector investment has been volatile and focused on certain countries and sectors.

Challenges to implementation of Agenda 21 objectives have also been identified at the local level (UN, 2002b), and financial barriers are the largest impediment there. Other factors include lack of support from national governments, difficulties affecting change in the policy sector (lack of empowerment), difficulties in generating community interest, and insufficient expertise and information. The last of these challenges reflects the urgent need for capacity-building to strengthen decision-making on economic, social, and environmental development in an integrated, place-based manner.

Successful implementation of Agenda 21 entails two kinds of transition. First, there remains a need to convert international obligations into national policy.[1] Second, this transition from normative standards to operational activities is usually associated with the creation of new knowledge (through scientific and technological research and the integration of indigenous or traditional knowledge) as part of a larger societal problem-solving process. Existing knowledge can also be applied in innovative ways. The need to reorient technology to respond to sustainability challenges was noted in *Our Common Future*.

> First, the capacity for technological innovation needs to be greatly enhanced in developing countries so that they can respond more effectively to the challenges of sustainable development. Second, the orientation of technology development must be changed to pay greater attention to environmental factors (WCED, 1987).

The relationship between environment and development means that sustainable development requires scientific input to decision-making as well as the application of technologies. Since 1992, science has been central to addressing Agenda 21 challenges in such areas as climate change, global warming, ozone depletion, water purity, land integrity, and air quality.

BOX 2-2
Corporate Roles in the Sustainability Transition

Some private enterprises have demonstrated their commitment to sustainable development by signing on to the U.N. global compact that promotes sustainable growth and good citizenship through corporate leadership. The compact calls on enterprises to adopt nine universal principles in the areas of human rights, labor standards, and the environment. Several hundred companies from all over the world have pledged their support to the compact and are implementing the nine principles. In addition, enterprises are securing certification of environmental management systems under the ISO 14000 standard of the International Organization for Standardization or the European Eco-Management and Audit Scheme. Rising stakeholder participation has led to initiatives including the global reporting initiative, which is developing a common framework for voluntary reporting on the economic, environmental, and social aspects of organization-level activities, products, and services.

SOURCE: UN (2002a).

IMPLEMENTING AGENDA 21 IN AFRICA

Efforts to implement Agenda 21 occur at continental, regional, national, and local levels. At the continental level the most ambitious activity is the establishment of the African Union, successor to the Organization of African Unity set up in 1963. Established in 2001, the African Union aims to enhance economic, political, and social integration and development of Africa; promote democracy; and resolve conflicts (NEPAD, 2001). To date, 37 of Africa's 53 countries have joined the union. Currently, the African Union is strengthening 14 regional integration communities to transform them into building blocks for economic coordination in Africa.

Africa's sustainable development aspirations are also reflected in continent-wide strategies of the New Partnership for Africa's Development (NEPAD, 2001). The NEPAD is a pledge by African leaders to develop natural and human re-

[1]Almost a decade after UNCED roughly 85 of the original 178 signatories of Agenda 21 have presented national policies.

sources. It recognizes that sustainability cannot be achieved until poverty is reduced, and living conditions are improved for the majority of the population. The NEPAD outlines the responsibility of African leaders to articulate national and regional priorities and to manage development by engaging people in their own development.

The Southern African Development Community (SADC), an intergovernmental group of 14 countries, is one example of a regional organization committed to equitable economic integration and sustainable development (Figure 2-2). In 2001 SADC completed an assessment of the regional implementation of Agenda 21 (SADC, 2001). Key issues addressed in the report are

- *Land resources.* Land-tenure issues, diminishing grazing lands, and land degradation have hindered progress on Agenda 21 implementation. However, several SADC countries have developed environmental management programs aimed at protecting and rehabilitating land resources.
- *Forest resources.* Rapid population growth and increased demand for food and energy resources are depleting forests. Groups have been organized to promote sustainable forestry management.
- *Water Resources.* Recurrent droughts, increasing water demands, and water pollution impede the implementation of Agenda 21. SADC water resources ministers have adopted a regional approach to integrate water resources development and management. This initiative has prevented conflicts over shared waters and improved access to potable water. SADC countries have also implemented reforms aimed at sustainable management of water resources.
- *Coastal zone.* Tourism and population pressure in coastal areas have thwarted implementation of Agenda 21. Critical to future implementation will be technology transfer, access to research findings and other information, and capacity-building for policy, institutional, and regulatory frameworks.

The Comité Permanent Inter-États de Lutte Contre la Sécheresse dans le Sahel (CILSS), an intergovernmental institution covering nine West African countries,[2] is a second regional organization concerned with sustainable development. CILSS has launched the Sahel 21 initiative that encourages Sahelians to define the future of development in their region.

Most countries in sub-Saharan Africa have encountered difficulties in gathering and analyzing national and sub-na-

[2]Burkina Faso, Cape Verde, Chad, Gambia, Guinea, Mali, Mauritania, Niger, Senegal.

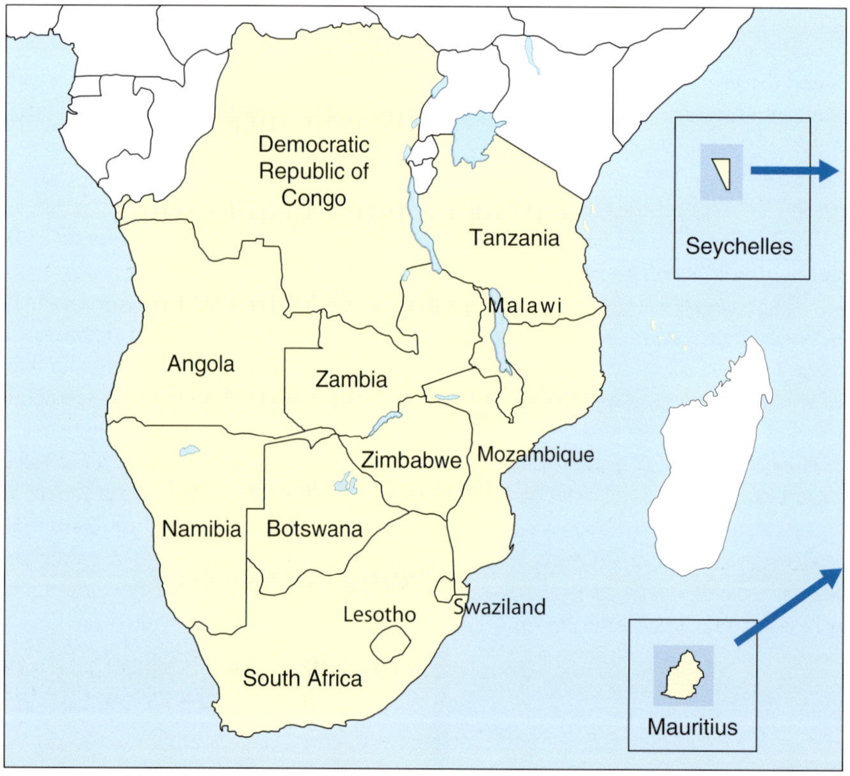

FIGURE 2-2 Map of the SADC member countries.

tional data on major Agenda 21 issues, such as natural resources inventories, climate change, and desertification. Thus, national-level reporting on these matters is uneven.

At the national level only South Africa has formally adopted a national Agenda 21 strategy (UN, 2002b). Although Burkina Faso, Cameroon, Cote d'Ivoire, and Ghana have not formalized national plans (NCSD, 2002), they have implemented national policies and laws that address issues related to sustainable development. For example, laws have been promulgated on environmental impact assessment; sustainable use of water, forests, and biodiversity; and management of solid wastes. In addition, over 95 percent of African countries have ratified the Rio conventions—the Convention on Biological Diversity, the U.N. Framework Convention on Climate Change, and the U.N. Convention to Combat Desertification (UNEP, 2001). At the local level there are LA21 activities in 28 African countries (Table 2-4).

African leaders discussed the challenges of implementing Agenda 21 at three regional consultations of the Rio+5 Forum[3]: North Africa and the Middle East (1996), West and Central Africa (1996), and South and East Africa (1997). The following challenges were identified at the consultations (Dorm-Adzobu, 2002):

- Inappropriate institutional frameworks in most countries, particularly a lack of coordination among ministries and across sectors;
- Inadequate coordination between governments, NGOs, and the private sector;
- Lack of appropriate legal frameworks;
- Lack of national consultation prior to signing international agreements and the proliferation of those agreements, which results in signing conventions without full knowledge of the implications to the countries and without having the capacity to translate these agreements into action;
- Poverty, illiteracy, and lack of awareness create problems in the development and implementation of sustainable development programs;
- Increasing gap between population growth and national economic output;
- Marginalization of women in the national development process; and
- Lack of vision and commitment by leaders to implement sustainable development.

Several documents (NEPAD, 2001; UNEP, 1999, 2001) provide detailed information on such topics as living conditions, national debt, armed conflicts, human health, environment, and education, giving a broad sense of environmental and developmental challenges facing African countries.

TABLE 2-4 African Countries with Local Agenda 21 Activities

Country	Number of LA21 Initiatives
Algeria	3
Benin	1
Burundi	2
Cameroon	1
Congo, Democratic Republic of	2
Egypt	7
Gabon	1
Ghana	3
Kenya	11
Libya	2
Madagascar	5
Mali	2
Malawi	4
Mauritania	1
Morocco	5
Mozambique	2
Namibia	5
Nigeria	5
Rwanda	1
Senegal	3
South Africa	20
Sudan	1
Tanzania	13
Togo	2
Tunisia	1
Uganda	5
Zambia	4
Zimbabwe	39

SOURCE: UN(2002b).

In summary, most of the progress in African countries toward implementation of Agenda 21 has been at the normative rather than the operational level, with the emergence of coordinating organizations such as SADC. In addition, national policies and laws relating to environment and development have been agreed upon and international environmental treaties have been signed by the majority of African countries. Effective translation of these measures into operational programs will require increased use of geographic information in decision-making (ECA, 2001). The integrative nature of GIS lends itself to a unified approach as opposed to the traditional fragmented approach to sustainable development (UN, 2002a). Application of GIS exemplifies the reorientation of technology in responding to sustainability challenges (WCED, 1987).

THE ROLE OF GEOGRAPHIC INFORMATION IN MEETING AGENDA 21 OBJECTIVES

The implementation of Agenda 21 has been slow (UN, 2002a) but the collection, analysis, and use of geographic information offers a starting point on the path to sustainable

[3]Rio+5 was held March 13-19, 1997, as a follow-up to the UNCED in 1992 in Rio de Janeiro. The purpose of the meeting was to review the progress of implementation of Agenda 21 and to move sustainable development from agenda to action.

development (Brooner, 2002). Society can benefit from the capabilities of geographic and other information sources and systems.[4] The NEPAD recognizes the value of these data and technologies: "[Information and Communications Technologies (ICTs)] can be helpful tools for a wide range of applications, such as remote sensing and environmental, agricultural, and infrastructural planning. In conflict management and control of pandemic diseases, ICTs will help towards the organization of efficient early warning mechanisms by providing tools for constant monitoring of tension spots (NEPAD, 2001)." In addition, the United Nations (2002a) notes the value of satellite remote-sensing systems as data sources for supporting sustainable development (Box 2-3).

Geographic data and information have the potential to play a role in the planning, implementation, and monitoring of many of Agenda 21's 38 action items (Table 1-1). In the committee's opinion these data and information are directly applicable to at least 20 of these action items (Table 2-5). These geographic data are often grouped into framework foundation data, framework thematic data, and other thematic data. Framework foundation data, such as geographic position and elevation, are central to most applications, whereas framework thematic data and other thematic data have specific applications, such as determining land ownership, possession, or use (Chapters 4, 5, and 6).

[4]Geographic information technologies (e.g., geographic information systems, global positioning systems, and cartographic, surveying, and remote-sensing technologies) are part of the broader spectrum of information and communications technologies that includes telephones, radios, and the Internet.

TABLE 2-5 Selected Agenda 21 Action Items (numbered by chapter from Agenda 21) and the Geographic Data to Address Them

	Framework Foundation Data (Fundamental layers of data used in many basic operations [Chapters 4 and 5])		
Agenda 21 Action Items	Position (Geodetic Control[a])	Aerial Image (Ortho-imagery[b])	Digital Elevation (Topography)
3. Combating poverty	✓	✓	✓
4. Changing consumption patterns	✓	✓	
5. Demographic dynamics and sustainability	✓	✓	✓
6. Protecting and promoting human health	✓	✓	✓
7. Sustainable human settlement development	✓	✓	✓
8. Integrating environment and development in decision-making	✓	✓	✓
9. Protecting the atmosphere			✓
10. Integrated planning and management of land resources	✓	✓	✓
11. Combating deforestation	✓	✓	✓
12. Managing fragile ecosystems: combating desertification and drought	✓	✓	✓
13. Managing fragile ecosystems: sustainable mountain development	✓	✓	✓
14. Promoting sustainable agriculture and rural development	✓	✓	✓
15. Conservation of biological diversity	✓	✓	✓
16. Environmentally sound management of biotechnology (DNA)			
17. Protecting oceans, seas, coastal areas, and rational use and development of living resources	✓ bathymetry	✓	✓ coastal
18. Protecting the quality and supply of freshwater resources	✓	✓	✓
19. Environmentally sound management of toxic chemicals	✓	✓	✓
20. Environmentally sound management of hazardous wastes	✓	✓	✓
21. Environmentally sound management of solid waste and sewage-related issues	✓	✓	✓
22. Environmentally sound management of radioactive wastes	✓	✓	✓

[a]Geodetic control is the common reference system for establishing the coordinate position (e.g., latitude, longitude, and elevation) of geographic data.

[b]An ortho-image is a specially processed image prepared from an aerial photograph or remotely sensed image that has the metric qualities of a traditional line map with the rich detail of an aerial image.

Temp = temperature; atmo = data on atmospheric conditions; precip = precipitation.

Agenda 21 (UNCED, 1992) specifies needs with respect to geographic data and information, and related technologies, including

- global sustainability indicators;
- data collection and use, including satellite-based remote-sensing;
- data assessment and analysis; and
- geographic information systems.

Additionally, Agenda 21 expressed the need for modern information frameworks and improved standards and methods for handling information, documentation about information, electronic networking capabilities, and partnerships among governments, international organizations, and the private sector. Many of these issues are fundamental to spatial data infrastructures (NRC, 1993; ECA, 2001; see also Chapter 4) that permit multisectoral, international, and transboundary data use and sharing. However, the data foundation on which this infrastructure can be built is far from complete or accessible to those that need it (Chapter 1), and the benefits of geographic data for decision-making on Agenda 21 issues have not been fully realized (ECA, 2001).

AN APPROACH TO EVALUATING THE ROLE OF GEOGRAPHIC INFORMATION IN SUSTAINABLE DEVELOPMENT APPLICATIONS

More than one approach can be taken to evaluate the use of geographic information for sustainable natural resource management and development in African countries and

	Selected Thematic Data (Supplemental layers of data that are often overlaid onto framework foundation data [Chapters 4, 5, and 6])									
Human Population (Total, Distribution, Change)	Land Ownership/ Possession (Cadastre)	Land Use/Land Cover			Biophysical Data (for Environmental Monitoring)					
		Local Scale	Regional Scale	Global Scale	Vegetation	Temp	Moisture Content	Precip	Water Quality	Atmo
✓	✓	✓	✓	✓	✓				✓	
✓	✓	✓	✓	✓	✓					
✓	✓	✓	✓	✓	✓				✓	
✓	✓	✓	✓	✓	✓				✓	✓
✓	✓	✓	✓	✓	✓			✓	✓	✓
✓	✓	✓	✓	✓	✓		✓	✓	✓	✓
✓	✓	✓	✓	✓	✓	✓	✓	✓	✓	✓
✓	✓	✓	✓	✓	✓		✓	✓	✓	✓
✓	✓	✓	✓	✓	✓	✓	✓	✓	✓	✓
✓	✓	✓	✓	✓	✓	✓	✓	✓	✓	✓
✓	✓	✓	✓	✓	✓	✓	✓	✓	✓	✓
✓	✓	✓	✓	✓	✓	✓	✓	✓	✓	✓
✓	✓	✓	✓	✓	✓	✓	✓	✓	✓	✓
✓		✓	✓		✓				✓	
✓	✓	✓	✓ coastal, oceanic	✓	✓	✓	✓	✓	✓	✓
✓	✓	✓	✓	✓	✓		✓	✓	✓	✓
✓	✓	✓	✓		✓	✓	✓	✓	✓	✓
✓	✓	✓	✓		✓	✓	✓	✓	✓	✓
✓	✓	✓	✓		✓	✓	✓	✓	✓	✓
✓	✓	✓	✓		✓				✓	✓

BOX 2-3
The Value of Earth Observation from Space

Earth observation from space is a valuable technological tool for understanding Earth. Such observations have provided long-term, consistent measurements of key variables for studying the state and variability of Earth's ecosystems. These observations provide a basis for rational action at global, regional, national, and local levels. Global phenomena, such as the greenhouse effect and El Niño, jostle for attention with regional problems, such as acid rain, and local problems, such as deforestation and soil erosion. Satellite remote-sensing has provided vital information on the environmental impacts, quantity and quality of resources, and inputs for integrated development planning for rural and urban areas. The launching of more than 230 instruments on more than 70 satellites over the next 10 to 15 years, with calibrated sensors providing a wide variety of data, will provide an opportunity for scientists to understand the complex interactions among various components of the Earth system. Parts of the world without the infrastructure to connect to the new global and information communication networks and people without access to education in the new technologies are being left behind.

SOURCE: UN (2002a, paragraph 162).

elsewhere. For example, in an issue-based approach each of the major natural resource or development applications is viewed separately. Major topics might include combating deforestation, conservation of biological diversity, health management, vulnerability to natural disasters, and combating poverty (e.g., Tables 1-1 and 2-5). However, most applications, especially those associated with natural resources, require approximately the same geographic information (e.g., a common coordinate system, elevation distribution, hydrologic data, land use or cover, political boundaries, transportation networks, population distribution), often collected multiple times to monitor change and evaluate impacts. Thus, the fundamental geographic information requirements repeatedly manifest themselves when marching through the issue applications (e.g., Table 2-5). Unfortunately, many developing countries lack the fundamental information for these applications.

Consequently, in order for developing countries in Africa and elsewhere to use geographic information to address sustainable development issues it is critical that (a) the importance of certain fundamental framework geographic information datasets is recognized, and (b) the current status and likely future availability of this geographic information is identified. This committee is not alone in recognizing the need for fundamental (or framework) geographic information for multiple applications. Many countries agree on the value of collecting and organizing certain geographic information within the structure of a national spatial data infrastructure (NRC, 1995, 2001; FGDC, 1997, 2002; ECA, 2001) (Chapter 4). The geographic information for a country may also be used to address global problems by adhering to Global Spatial Data Infrastructure protocols (GSDI, 2002). Once these most important geographic information layers are available they may be used many times for many different applications, ideally within rigorous decision-support systems. Thus, these data become an important national and international asset (ECA, 2001) and are the baseline for future analyses and assessments.

The following comparison demonstrates why the committee focused on assessing the major types of geographic information required by many applications rather than adopting an issue-by-issue approach. Consider the problem of famine. To predict famine and make plans to minimize its impact on humans it is useful to have information on land cover (including crop type, biomass, and leaf-area-index), precipitation (recent and predicted), soil moisture (recent and predicted), population distribution, transportation infrastructure (to deliver food relief or relocate people), and political boundaries (to help address politically sensitive issues). These types of data are used in famine decision-support systems such as the Famine Early Warning System (Chapters 3 and 7).

Similarly, consider deforestation and biodiversity assessment (e.g., Box 2-4). Both issue applications require information including population distribution (to identify the likelihood for forest-cover or biodiversity changes due to selective harvesting of forest resources or other agricultural practices), land use and cadastral information, land cover (to reveal the historical and current geographic distribution of rangeland or forest cover, in part as a surrogate for habitat in biodiversity applications), elevation distribution (to provide insight into watershed runoff, soil erosion, and habitat carrying capacity), and transportation infrastructure (to indicate new tertiary roads that provide access to forests and enable deforestation and habitat fragmentation). When these types of geographic information are modeled in conjunction with in situ faunal information it becomes possible to assess biodiversity (Savitsky and Lacher, 1998). Subsequently, biodiversity information can be used to perform "gap analysis" for wise use of financial resources to protect fragile, biologically diverse habitats (Edwards et al., 1995).

It is evident from the overlap in data needs to address biodiversity, deforestation, and famine that the same fundamental geographic information may be used in more than one issue application. Thus, this report concentrates on fundamental geographic data (Chapters 5 and 6) that provide

BOX 2-4
Biodiversity in Africa

Biodiversity is the biological diversity in an environment as indicated by numbers of different species of plants and animals (Merriam-Webster, 1994). People everywhere depend on functioning natural ecosystems for their survival. Population growth, the clearing of wildlands, and overuse of pesticides and fertilizers are among the principal causes of loss of biodiversity in developing countries (Wilson, 1992). Poverty, social conflict, and disease are among the results of environmental despoliation.

The economic and social benefits of biodiversity range from discovering new sources of pharmaceuticals to enhancing tourism. Coastal areas in Kenya and Tanzania (Chapter 8) gain economically from aquaculture, tourism, and disaster protection. Conservancies in Namibia also benefit through tourism. Their use of geographic information for wildlife management (Chapter 7) promotes equitable land use and tourism potential.

Geographic information can be used to support African governments' efforts to monitor the state of natural resources for environmental protection, hazard reduction, and for international environmental reporting required by treaties such as the Convention on Biological Diversity and the Convention on International Trade in Endangered Species of which many African countries are signatories.

The Eastern and Southern Africa Regional Biodiversity Forum convened representatives from the public, private, and non-profit sectors in Mombasa, Kenya, in 2000 to discuss using biodiversity to strengthen livelihoods (Emerton and Maganya, 2000). Representatives from more than 20 countries identified four major conclusions:

- Rural people are ultimately the custodians of Africa's rich biodiversity;
- Investment from private industry offers a valuable, underused means of protecting biodiversity and promoting the equitable sharing of resources;
- There is urgent need for better cooperation between the public and private sectors on biotechnology research in agriculture; and
- National action plans should be created to address these concerns.

Geographic information including social, economic, and environmental data from the local, national, and regional levels can support these goals.

SOURCE: Emerton and Maganya (2000)

the geographic foundation for many Agenda 21 issues and on the decision-support systems and capacity-building efforts needed to link these data to their applications.

SUMMARY

Agenda 21 is being implemented worldwide at all scales and in all sectors, but progress has not been as rapid as anticipated at the UNCED in 1992. The use of geographic data and related technologies will help overcome a number of implementation challenges, in particular the traditional fragmented approach toward sustainable development. In Africa, organizational frameworks and coordination will enhance the use of these data and technologies. The next chapter illustrates activities in Africa that are already using geographic data.

REFERENCES

Brooner, W. G. 2002. Promoting Sustainable Development with Advanced Geospatial Technologies. Photogrammetric Engineering and Remote Sensing 68(3):198-205.

Dernbach, J. C., 2002. Sustainable Development: Now More Than Ever. Environmental Law Reporter 32 (1):10003-10019.

Dorm-Adzobu, C., compiler. 2002. Synthesis Report of Africa and the Middle East Consultations on Implementation of Sustainable Development. Available at <http://www.ecouncil.ac.cr/rio/regional/africa/afmesynt.htm>. Accessed June 25, 2002.

ECA (United Nations Economic Commission on Africa). 2001. The Future Orientation of Geoinformation Activities in Africa: A Position Paper. Available at <http:// www.uneca.org/eca_programmes/it_for_ development/geoinfo/ FutureGIAfrica.PDF>. Accessed June 25, 2002.

Edwards, T. C., C.G. Homer, S.C. Bassett, A. Falconer, R.D. Ramsey, and D. W. Wighe. 1995. Utah Gap Analysis: An Environmental Information System. Logan, Utah: Utah Cooperative Fish and Wildlife Research Unit.

Emerton, L. and J. Maganya (eds.). 2000. Report of the First Eastern and Southern Africa Regional Biodiversity Forum: Using Biodiversity to Strengthen Livelihoods. Mombasa, Kenya: IUCN – The World Conservation Union, East Africa Regional Office, Nairobi. Available at <http://www.undp.org/bpsp/regional/docs/E_S_Africa_RBF_Feb2000.pdf>. Accessed July 8, 2002.

FGDC (Federal Geographic Data Committee). 1997. Framework: Introduction and Guide, Washington, DC: Federal Geographic Data Committee.

FGDC. 2002. Overview: What the Framework Approach Involves. Available at <http://www.fgdc.gov/framework/overview.html>. Accessed April 24, 2002.

GSDI (Global Spatial Data Infrastructure). 2002. Developing Spatial Data Infrastructure: The SDI Cookbook. D. Nebert, ed. Available at <http:// www.gsdi.org/pubs/cookbook/cookbook0515.pdf >. Accessed June 28, 2002.

Juma, C. 2002. The global sustainability challenge: from agreement to action. International Journal of Environmental Issues 2(1/2):1-14.

Lake, R.W. 2000. Contradictions at the local scale: local implementation of Agenda 21 in the USA. Pp. 70-90 in Consuming Cities: The Urban Environment in the Global Economy after the Rio Declaration, N. Low, B. Bleeson, I. Elander, and R. Lidskog, eds. London: Routledge.

Merriam-Webster. 1994. Merriam-Webster Pocket Dictionary. Merriam-Webster: Springfield, MA.

NCSD (National Councils for Sustainable Development). 2002. NCSD Knowledge Network. Available at <http:// www.ncsdnetwork.org/ncsd/main.cfm>. Accessed May 20, 2002.

NEPAD (New Partnership for Africa's Development). 2001. The New Partnership for Africa's Development. Available at <http://www.dfa.gov.za/events/nepad.pdf>. Accessed May 29, 2002.

NRC (National Research Council). 1993. Toward a Coordinated Spatial Data Infrastructure for the Nation. Washington, DC: National Academy Press.

NRC. 1995. A Data Foundation for the National Spatial Data Infrastructure. Washington, DC: National Academy Press.

NRC. 1999. Our Common Journey: A Transition Toward Sustainability. Washington, DC: National Academy Press.

NRC. 2001. National Spatial Data Infrastructure Partnership Programs: Rethinking the Focus. Washington, DC: National Academy Press.

SADC (Southern African Development Community). 2001. SADC Progress Report on the Implementation of Agenda 21 and Sustainable Development: A Report Submitted to the 2002 World Summit on Sustainable Development. Available at <http://www.johannesburgsummit.org/web_pages/sadc_prepcom_progress_report.pdf>. Accessed May 20, 2002.

Savitsky, B. G. and T. E. Lacher, Jr. 1998. GIS Methodologies for Developing Conservation Strategies. New York: Columbia University Press.

UN (United Nations). 2002a. U.N.'s Rio Plus 10 Review from Prepcom2. Available at <http://www.johannesburgsummit.org/html/documents/no170793sgreport.pdf>. Accessed May 29, 2002.

UN. 2002b. Second Local Agenda 21 Survey, UN Department of Social and Economic Affairs, background paper No. 15 (DESA/DSD/PC2/BP15). Available at <http://www.iclei.org/rioplusten final_document.pdf>. Accessed April 14, 2002.

UNCED (United Nations Conference on Environment and Development). 1992. Report of the United Nations Conference on Environment and Development, Rio de Janeiro, 3-14 June 1992. Rio de Janeiro: UN. (Annex I: Rio Declaration; Annex II: Agenda 21)

UNEP (United Nations Environment Programme). 1999. Global Environmental Outlook 2000. London: Earthscan.

UNEP. 2001. Assessment of Progress on Sustainable Development in Africa Since Rio (1992). Available at <http://www.ecs.co.sz/wssd/roa_assessment_of_progress.htm>. Accessed June 25, 2002.

WBCSD (World Business Council for Sustainable Development) website. 2002. Case studies. Available at <http://www.wbcsd.org/casestud/index.htm>. Accessed May 20, 2002.

WCED (World Commission on Environment and Development). 1987. Our Common Future. New York: Oxford University Press.

Wilson, E. O. 1992. The Diversity of Life. New York: W.W. Norton and Company.

3

Geographic Information Activities in Africa

INTRODUCTION

Africa has a small but growing community of data providers, data processors and analysts, trainers, technicians, data and information users (decision-makers at many levels), and advocates. The community is becoming increasingly coordinated, and consists of Africans and international partners from non-governmental organizations, universities, private companies, and foreign governments including the space and aid agencies that currently are a major source of geographic data, information, training, and support.

This chapter introduces some geographic information activities in Africa and highlights the applications in tables. It focuses on activities in regions of Africa targeted by the U.S. government's public-private partnership called Geographic Information for Sustainable Development (GISD). These regions are the Upper Niger basin in West Africa, the Limpopo-Zambezi region of southeastern Africa, and two in East Africa (the African Great Lakes region and the Kenya/Tanzania coastal zone) (Figure 3-1). The chapter presents examples of ongoing activities and enumerates the range of applications rather than evaluating individual activities.

ORGANIZATIONS WITH CONTINENT-WIDE ACTIVITIES

This section describes five organizations with continent-wide activities that utilize or promote collection and analysis of geographic data. These are

(1) the Famine Early Warning System Network (FEWS NET),
(2) U. N. Economic Commission for Africa (ECA),
(3) U. N. Food and Agriculture Organization (FAO),
(4) Environmental Information Systems-Africa (EIS-Africa), and
(5) the Collaborative Research Support Programs (CRSP).

Famine Early Warning System Network

FEWS NET (Chapters 6 and 7) was initiated by the U.S. Agency for International Development (USAID) in 1980. This activity uses satellite imagery[1] to provide an estimate of the amount and vigor of vegetation across Africa. FEWS NET is the principal food security information-based approach to preparedness and planning in sub-Saharan Africa, and aims to empower Africans to find solutions to food insecurity problems (FEWS NET, 2002).

Seventeen African countries participate in the FEWS NET.[2] The network is supported by implementation partners[3] and network partners (Centre Regional AGRHYMET[4] [Niamey, Niger], Drought Monitoring Centre [Nairobi, Kenya], and SADC Regional Early Warning Unit [Harare, Zimbabwe]), which provide imagery, analyses, and reports on vegetation conditions.

Economic Commission for Africa

ECA promotes the use of geographic information in Africa. Resources developed by ECA include the *Inventory of Education and Training Facilities in African GIS*, the *Cartographic Inventory Atlas for Africa*, and the *Database on the Status of Cartographic Coverage and Programs in Africa*. ECA supports regional Centers (e.g., Nairobi) through its

[1]From NOAA's Advanced Very High Resolution Radiometer (AVHRR). Images collected from 1972 through the present show where abnormal vegetation conditions exist within Africa at a given time.

[2]Burkina Faso, Chad, Eritrea, Ethiopia, Kenya, Malawi, Mali, Mauritania, Mozambique, Niger, Rwanda, Somalia, (southern) Sudan, Tanzania, Uganda, Zambia, and Zimbabwe.

[3]Chemonics International, NASA, NOAA, U.S. Department of Agriculture.

[4]AGRHYMET stands for AGRiculture, HYdrology, METeorology

FIGURE 3-1 Case-study regions in Africa: the Upper Niger basin in West Africa, the Limpopo-Zambezi region of southeastern Africa, and two in East Africa (the African Great Lakes region and the Kenya/Tanzania coastal zone). SOURCE: Dan Zimble, ESRI.

geographic information activities at ECA headquarters in Addis Ababa. Harnessing Information for Development is one of six core programs of ECA, and geographic information is a part of the Development Information Services Division (DISD). In addition to supporting the regional centers and the five other core program areas of ECA, DISD activities include advisory services, developing inventories and databases, and organizing conferences and meetings. Since 1963 the ECA has organized a number of U.N. Cartographic Conferences, which in 1999 became part of the activities of the Committee on Development Information (CODI). This committee provides policy and technical guidance for the implementation of the core program on information for development, and these recommendations are passed on to the Conference of African Ministers of Planning and Development.

Food and Agriculture Organization

In 1995, the FAO initiated the Africover project (Chapter 6) in response to national requests for assistance in obtaining reliable, geographically referenced information on natural resources for use in early warning systems, forest and rangeland monitoring, catchment management, and biodiversity or climate-change studies at national and regional levels. Ten countries of the eastern African sub-region[5] are involved in the first application of this initiative. These countries are creating a digital geographically referenced land-cover database, the Multipurpose Africover Database for Environmental Resources (FAO, 2002). Funding for the project is from the Italian and U.S. governments.

Environment Information Systems-Africa

With goals similar to ECA, EIS-Africa (Chapter 4) is a network of public- and private-sector institutions and experts. Founded in 1999, it promotes "access to and use of environmental information in the sustainable development process."[6] EIS-Africa, currently with 24 member states, is the outgrowth of an earlier EIS project. The organization has many activities, including a bi-annual conference series called Africa GIS.[7] The most recent Africa GIS conference in Nairobi, Kenya, in November of 2001 attracted 450 attendees, the largest number to date.[8] EIS-Africa is a nonprofit pan-African organization funded and supported by several donors and international development agencies, including the World Bank, the Government of Norway, USAID, German Agency for Technical Cooperation, and several U.N. agencies including the U.N. Development Program (UNDP), and the U.N. Environment Program (UNEP).

Collaborative Research Support Programs

CRSPs promote agricultural research within Africa and elsewhere (Chapter 7). Created by USAID and the Board for International Food and Agriculture Development in 1975, CRSPs employ U.S. land grant universities to carry out the international food and agricultural research mandate of the U.S. government. CRSPs consist of universities working with developing-country national agricultural research systems, international agricultural research centers, U.S. agribusiness, private voluntary organizations, developing-country colleges and universities, private agencies, USAID/Washington and USAID missions, and other U.S. federal agencies such as the U.S. Department of Agriculture (CRSP, 2002). Presently, there are nine CRSPs, and all of them have activities in Africa (Table 3-1).

[5]Burundi, Democratic Republic of Congo, Egypt, Eritrea, Kenya, Rwanda, Somalia, Sudan, Tanzania, and Uganda.

[6]A. Bassolé, EIS-Africa, personal communication, 2002.

[7]UNITAR (U.N. Institute for Training and Research), and UNEP (U.N. Environment Programme) also are involved with Africa GIS.

[8]D. Tunstall, World Resources Institute, personal communication, 2002.

GEOGRAPHIC INFORMATION ACTIVITIES IN CASE-STUDY REGIONS

West Africa

The diverse populations of the Upper Niger River region (which includes the Sahelian ecological zone) (Figure 3-1) depend on the natural resources of the region's arid ecosystems. Declines in land productivity, owing in part to desertification and climate change, have increased the vulnerability of people to food insecurity, especially over the last three decades (GISD, 2001). Table 3-2 provides a sample of some of the organizations that are using geographic data to address the region's environmental challenges.

Featured Activities

Among the prominent, internationally supported centers in the region is AGRHYMET in Niamey, Niger. AGRHYMET uses remotely sensed data and maps to address natural resource management and food security issues for the nine West African member states of the Permanent Interstate Committee to Combat Drought in the Sahel/Comité Permanent Inter-´Etats de Lutte Contre la Sécheresse dans le Sahel (CILSS).[9] AGRHYMET is a focal point for FEWS NET activities (Chapter 6) and is a center of training and educational outreach in West Africa. Established in 1974, the center received financial assistance initially from USAID and later from Belgium, Denmark, France, and Italy, the Netherlands, and the World Meteorological Organization.

The Centre of Applied Research in Population and Development (CERPOD) (Chapter 5) is part of the Institute of the Sahel in Bamako, Mali. CERPOD concentrates on research and applied issues relating to population and development. In its work CERPOD integrates socioeconomic data with environmental data. Member countries from CILSS provide partial support for CERPOD, in addition to USAID, Canadian International Development Agency, and the U.N. Population Fund.

East Africa

In East Africa (Figure 3-1) food insecurity is a long-standing chronic issue (LEWS, 2001). Factors that have led to this situation include the region's weather and climate variation, increasing human population, political instability, and land-use/cover changes. Indeed, in the late 1980s, an estimated 71 million people, or 46 per cent of the population in the Greater Horn of Africa, were reported to be chronically food insecure (LEWS, 2001).

In addition to food insecurity, the Great Lakes of East

[9]Burkina Faso, Cape Verde, Chad, Gambia, Guinea, Mali, Mauritania, Niger, Senegal.

TABLE 3-1 A Summary of Applications of Geographic Data by Collaborative Research Support Programs (CRSPs)

Name of CRSP, Year Established, Website	Participating African Nations	Applications of Geographic Data
International Corn and Millet (INTSORMIL), 1979 <http://intsormil.unl.edu/>	Burkina Faso, Chad, Ghana, Mali, Niger, Nigeria, Senegal	Soil and rainfall monitoring; tracking population and crop density to ensure proper food distribution, educating farmers to increase production
Bean/Cowpea, 1980 <http://eastafricacrsp.wsu.edu/> <http://www.isp.msu.edu/CRSP/crsphc.htm> <http://crsps.unl.edu/virttour.html>	Egypt, Jordan, Malawi, Nigeria, Senegal, Tanzania, Zimbabwe	Drought, insect, and disease monitoring to improve crop yields
Peanut, 1982 <http://www.griffin.peachnet.edu/pnutcrsp.html>	Angola, Benin, Botswana, Burkina Faso, Ghana, Senegal, Malawi	Simulation of crop systems to improve production efficiency and natural resource management; monitoring nitrogen levels in soil
Global Livestock, 1978 <http://cnrit.tamu.edu/aflews/> <http://glcrsp.ucdavis.edu/txt_index.html>	Eritrea, Ethiopia, Kenya, Tanzania, Uganda,	Livestock monitoring; monitoring the effects of production on the environment; natural resource conservation and management
Pond Dynamics/Aquaculture (PD/A), 1982 <http://pdacrsp.orst.edu/projects_people/crsp_intro.html>	Egypt, Kenya, Rwanda	Identification and monitoring of constraints to aquaculture; tracking physical, chemical, and biological processes of aquatic ecosystems
Integrated Pest Management (IPM), 1993 <http://www.ag.vt.edu/ipmcrsp/>	Ethiopia, Ghana, Kenya, Mali, Senegal, Uganda, Zimbabwe	Research and education programs to manage plant diseases and pests; development of Africa IPM Link for information exchange and electronic communication
Sustainable Agriculture and Natural Resources Management (SANREM), 1992 <http://www.oird.vt.edu/sanremcrsp/sanremcrsp.html>	Mali	Enhancing decision-making capacity (by providing appropriate data, information, tools and methods of analysis) to improve the sustainability of natural resources
Broadening Access and Strengthening Input Systems (BASIS), 1996 <http://aem.cornell.edu/special_programs/AFSNRM/Basis/index.html> <http://www.basis.wisc.edu/history.html>	Kenya, Madagascar	Monitoring natural resource depletion and degradation of soils; assessing interactions between water, land, labor and financial markets and policy
Soil Management (SM), 1981 <http://www.tradeoffs.montana.edu>	Ethiopia, Ghana, Malawi, Mali, Niger, Senegal, Uganda, Zambia	Assessing outcomes of alternative soil management practices; soil monitoring (pesticide leaching, erosion, and fertility decline)

Africa are facing other human and environmental challenges, including

- invasive species, such as water hyacinth that contributes to fisheries losses;
- water and soil contamination;
- pressure on rare wetlands and montane forests;
- soil erosion in the densely settled East African Highlands;
- ethnic and herder/farmer conflicts; and
- human and animal health problems, such as schistosomiasis, Rift Valley fever, and malaria, which are often closely linked to irrigation, land-use change, and water resource use.

The coastal region of East Africa, in Kenya and Tanzania, is undergoing notable changes. These changes include intensification of agriculture and mariculture, increasing tourism, rapid expansion of coastal cities, especially Mombasa and Dar es Salaam, and increasing population through natural increase and in-migration. Pollution and over-fishing threaten important coral reefs and fish-breeding grounds from Lamu, Kenya, to Mtwara in southern Tanzania.

Geographic data illuminate most of the human and environmental issues in East Africa. Table 3-3 lists some of the organizations and programs that use geographic data in this region.

TABLE 3-2 West African Organizations, Programs, and Activities Using Geographic Data

Organization	Applications of Geographic Data
National Geographic Institute of Benin	Mapping; topography
National Centre of Remote Sensing (CENATEL) (Benin)	Vegetation monitoring; environment
Geographic Institute of Burkina Faso	Mapping; topography
National Program for Environment Information Management (PNGIM) (Burkina Faso)	Vegetation monitoring; environment
National Institute for Agronomic Research (Burkina Faso)	Agriculture; natural resources
Centre de SIG et de Télédétection Adjaratou (Burkina Faso)	Training; GIS services
Ecole Inter-Etats des Ingenieurs de l'Equipement Rural (Burkina Faso)	Training; GIS services
Famine Early Warning System Network (FEWS NET) (Burkina Faso, Chad, Mali, Mauritania, Niger, Senegal)	Famine early warning
Centre for Cartography and Remote Sensing (CCT) (Cote d'Ivoire)	Mapping
Comité National de Télédétection et d'Information Géographique (CNTIG) (Cote d'Ivoire)	Applying remote-sensing and GIS technology in all economic development sectors in Côte d'Ivoire (under the office of the Prime Minister)
Institut de Géographie Tropicale (IGT) at the University of Cocody in Abidjan	Training in remote-sensing and GIS
Centre Universitaire de Recherche Appliquée en Télédétection (CURAT) at the University of Cocody in Abidjan	Training in remote-sensing and GIS
Division of Water Resources (Gambia)	Water resources
National Environment Agency (NEA) (Gambia)	Vegetation monitoring; environment
Centre for Remote Sensing and Geographic Information (CERSGIS) (Ghana)	Training; research; varied applications
Environment Protection Agency (Ghana)	Environment
Sambus Company Limited (Ghana)	Training; commerce
Upper Niger Hydro-Electrical Management Project (Guinea)	Water resources
National Forestry and Waters Service) (Guinea)	Forestry; environment
National Meteorological Service (Guinea Bissau)	Climatology
Direction General of Natural Resources (Guinea Bissau)	Water resources
Sahel Institute (INSAH) (Mali)	Demography; agronomy; research
Rural Economy Institute (IER) (Mali)	Agronomy; natural resources management; research
International Crop Research Institute for the Semi-Arid Tropics (ICRISAT) (Mali)	Agronomy; research
Agrometeorology Service (Mauritania)	Agrometeorology
Regional Centre for Training in Aerospace Survey (RECTAS) (Nigeria)[a]	Training; mapping
University of Lagos, Laboratory for Cartography and Remote Sensing (Nigeria)	Training; research; mapping
Ecological Monitoring Centre (CSE) (Senegal)	Natural resources; livestock monitoring
Senegal Agronomic Research Institute	Agronomy; research
Interstate School for Veterinarian Sciences (Senegal)	Livestock monitoring
National Center for Research Support (Chad)	Research; mapping; agriculture
Farcha's Laboratory for Veterinarian Research (Chad)	Livestock monitoring
Water Resources and Meteorological Service	Water resources, agrometeorology
University of Benin, Department of Geography (Togo)	Training; research
AGRHYMET Regional Centre (Niger)	Early warning; crop production forecasts; natural resources; environment
National Geographic Institute (Niger)	Mapping; topography
Environment Technical Agency (Niger)	Forestry; environment
Geographic Information System of Niger	Water resources; hydraulics
Sequestration of Carbon in Soil Organic Matter (SOCSOM) (Senegal)	Climate change; agriculture; rural development
Operation Haute Vallee du Niger (OHVN) (Mali)	Agriculture; natural resource management
U.N. Environment Programme/ Réseau pour l'environnement et le développement durable en Afrique (UNEP/REDDA)	Environmental management

[a]Discussed in Chapter 8.

SOURCES: A. Bassolé, EIS-Africa, personal communication, 2002; J. Oliveira, Mitchell Group, personal communication, 2002; <http://www.opengis.org/gisd>.

TABLE 3-3 East African Organizations, Programs, and Activities Using Geographic Data

Organization	Applications of Geographic Data
Kenya Institute of Surveying and Mapping	Training; mapping products
Kenya Agricultural Research Institute (KARI)	Agricultural and livestock research; resource management
Kenya Soil Survey	Research: soil science
Department of Resource Surveys and Remote Sensing (Ministry of Planning and National Development, Kenya)	Land use; land cover; land degradation; drought; Early Warning Systems forestry and vegetation; livestock; wildlife; crop production forecasting
Kenya Wildlife Service	Research; training; education; conserve, protect, and manage biodiversity
Consultive Group on International Agricultural Research (CGIAR)/ International Center for Research in Agroforestry (ICRAF)/ International Livestock Research Institute (ILRI) (Kenya)	Research: crop productivity, forestry and agroforestry, water management, aquaculture, and livestock; training; education; information services research and training in natural resource management, land degradation, vegetation change
Kenya Polytechnic POLYGIS Centre	Training
Kenya Marine and Fisheries Research Institute (KEMFRI)	Research: fisheries; environmental and ecological studies; chemical and physical oceanography
Regional Centre for Mapping of Resources for Development (RCMRD) (Kenya)[a]	Natural resource development and environmental management; training and services
International Union for the Conservation of Nature and Natural Resources/ Species Survival Commission African Elephant Specialist Group (IUCN/SSC) (Kenya)	Conservation and management of African elephants
National Environmental Management Authority (Uganda)	Environmental management; monitoring, planning and coordination of environmental matters (Under the Ministry of Water, Lands, and Environment)
Uganda Land Commission	Monitoring, management and protection of lands (Under the Ministry of Water, Lands, and Environment)
International Center for Living Aquatic Resources (ICLARM) (Egypt, Malawi)	Research: food and environment; collaboration of farmers, scientists, and policy-makers to help maintain environment and alleviate poverty
African Highlands Initiative (AHI)	Natural resource management; agriculture
Global International Waters Assessment (GIWA) (East African Rift Valley Lakes)	Water resources; environmental problems and threats to transboundary water bodies
GIS-Network on Snailborne Infections with Special Reference to Schistosomiasis (GNOSIS-GIS) (Great Lakes/Horn of Africa)	Mapping for medical research (schistosomiasis)
Tanzania Coastal Management Partnership with National Environmental Management Council (TCMP with NEMC)	Coastal change detection
UNEP-DEWA (Division of Early Warning and Assessment)	Environmental data for decision-making; sustainable human development
UNEP-GRID (Global Resource Information Database) (Nairobi)	Environmental data and information for decision-making; supports environmental research and information networks in Africa, the Mediterranean region, and West Asia; provides advisory services to other UNEP units on data management, GIS, remote sensing and environmental information systems

[a]Discussed in Chapter 8.

SOURCE: <http://www.icraf.cgiar.org/about/about.htm>, <http://www.opengis.org/gisd>; <http://www.unep.org/unep/eia/ein/grid/web/document/grid.htm>; comments from contributors (Appendix B).

Featured Activities

Established in 1997, the Integrated Coastal Management project in Tanzania and Kenya (Chapter 8) is part of the U.S. GISD initiative. This project uses Landsat satellite data and applies GIS technologies for coastal management planning and decision-making. The project measures rates of change in resource and land-use patterns and identifies priority locations for coastal-action planning, special-area management, aquaculture siting, tourism-development planning, and land-use zoning.

The Regional Centre for Mapping of Resources for Development in Nairobi (Chapter 8), one of ECA's regional centers, was established in 1975. The original objectives of the center concentrated on training, technical and consulting services, and advisory services to member states.[10] Changes in technology, capacity in the region it serves (the center has trained 3,000 Africans), and social and operational needs of member countries (now totaling 25 African countries) have resulted in re-focusing of RCMRD's programs to address natural resource development and environmental management challenges.

Originally funded by USAID, the RCMRD now operates with funds from contracting member states, and from donors that include USAID, U.N. Development Programme (UNDP), FAO, World Bank, International Development Research Center (IDRC), United Nations Educational, Scientific, and Cultural Organization (UNESCO), Bank for Economic Development in Africa (BADEA), and a number of bilateral donors such as France, India, Italy, and the Netherlands. The center also generates its own funds from technical services such as training, aerial photography, servicing of survey and mapping equipment, and consulting in natural resources surveys and mapping (RCMRD, 2002).

The Livestock Early Warning System (LEWS) (Chapter 7) monitors nutrition and livestock health, and the food security of people in East Africa. It applies integrated remotely sensed weather data, point-based biophysical modeling, and geographic data on animal and vegetation distribution to serve decision-makers at many levels. LEWS provides timely information on projected trends in livestock condition (e.g., weight, mortality, milk, reproduction), forage supply, and crop stability—an estimated six to eight weeks earlier than other monitoring systems. LEWS is funded by USAID and is implemented by Texas A&M University.

The Global Access Initiative[11] in Uganda (Chapter 8) is part of a coalition of non-governmental organizations (NGOs) from developed and developing countries. The purpose of the initiative is to ensure that civil society has in place all the critical elements of an access system (information, participation, and justice). The initiative seeks common standards for information access within and between countries.

The Limpopo/Zambezi Region of Southern Africa

The Limpopo and Zambezi river basins lie within parts of Mozambique, Zimbabwe, South Africa, Namibia, Angola, and Zambia (Figure 3-1). These river basins are prone to devastating floods and famine. Other challenges within the region include natural resource management, settlement changes and their impacts on agricultural and environmental conditions, and conflict over land and water resources. Potential and existing applications of geographic information in this region focus on natural disaster relief, community-based and transboundary natural resource management, interactions between land uses and economic development, river-basin management, flood control, and infectious disease transmission (GISD, 2001). Table 3-4 lists some of the organizations and programs that use geographic data in this region.

Featured Activities

The Miombo Network (Chapter 7) examines the rates and causes of land-cover changes (including prevalent fires), predicting the consequences of land-use and land-cover changes, developing a predictive understanding of Miombo woodland structure and function; and understanding the determinants of the distribution of species and ecosystems. The Miombo Network operates under the auspices of the IGBP/IHDP (International Human Dimensions Programme on Global Environmental Change/ International Geosphere-Biosphere Programme) Land Use and Cover Change (LUCC) Project and the IHDP/IGBP/WCRP[12] Global Change System for Analysis, Research and Training (START). Miombo members in Africa include government, university and research institutions in the Democratic Republic of the Congo, Malawi, Mozambique, Tanzania, Zambia, and Zimbabwe. The network also includes universities, research institutions and NGOs in Europe, Australia, and the United States.

The Community Based Natural Resource Management (CBNRM) (Chapter 8) program in Namibia applies geographic data for mapping conservancy boundaries, land-use planning, monitoring, and communication and training. The program was initiated through a partnership among the Namibian Ministry of Environment and Tourism and NGOs and donors that include USAID, World Wildlife Fund, Department for International Development, CIDA, Namibia Nature Foundation, Endangered Wildlife Trust, Canadian Ambassador's Fund, Hivos, and Ford Foundation.

[10] Kenya, Tanzania, Uganda, Zambia.

[11] Hosted by the World Resources Institute in Washington, D.C.

[12] WCRP = World Climate Research Programme

TABLE 3-4 Southern African Organizations, Programs, and Activities Using Geographic Data

Organization	Applications of Geographic Data
African Wildlife Foundation	Wildlife conservation; habitat protection; landscape ecology; community-based natural resource management
Miombo Network	Land-use and land-cover change detection
Okavango Wildlife Society	Wildlife and ecological management
Program for Regional Information Sharing and Management on Environment and Sustainable Development (PRISMES), World Bank	Natural resource management
Community Based Natural Resource Management (CBNRM) Program (Namibia)	Ecosystem management; wildlife
Geologic Survey (Namibia)	Mineral resources; geological engineering; land-use planning
University of Namibia, Department of Geography and Environmental Studies	Physical and human geography; environmental planning; regional studies
Regional Centre for Southern Africa	Natural resources management; agriculture
World Wildlife Foundation's Living in a Finite Environment Programme (Namibia)	Natural resource management
World Wildlife Fund's Communal Areas Management Programme for Indigenous Resources (CAMPFIRE Programme) (Zimbabwe)	Natural resource management
Food, Agriculture and Natural Resources (FANR) Development Unit (Zimbabwe)	Food security; natural resource management; early warning; vulnerability assessment
Desert Research Foundation of Namibia	Desertification; ecosystem function and structure
National Forest Research Center (Namibia)	Desertification; forestry; remote sensing; horticulture
SAFARI 2000	Land-atmosphere processes
Southern African Fire Network (SAFNET)	Monitoring burned areas (natural and anthropogenic fires)
Flood Hazard Monitoring (Mozambique)	Flood prediction models; rainfall analysis
Institute for Fisheries Management and Coastal Community Development (IFM)	Fisheries management and development

SUMMARY

Africa's geographic information community is diverse and growing. This chapter highlights some of the applications of geographic data to environmental and developmental challenges. The next chapter explores how these diverse activities can be linked together through spatial data and telecommunications infrastructures.

REFERENCES

CRSP (Cooperative Research Support Programs). 2002. Historical Perspectives and U.S. Government Policy. Available at <http://crsps.unl.edu/whycrsps.htm>. Accessed May 7, 2002.

FAO (Food and Agriculture Organization). 2002. Available at <http://www.fao.org/>. Accessed May 20, 2002.

FEWS NET (Famine Early Warning System Network). 2002. Available at <http://www.fews.net/>. Accessed May 20, 2002.

GISD (Geographic Information for Sustainable Development). 2001. Available at <http://www.opengis.org/gisd/> Accessed May 20, 2002.

LEWS (Livestock Early Warning System). 2001. Available at <http://cnrit.tamu.edu/lews/description.html> Accessed May 20, 2002.

RCMRD (Regional Centre for Mapping of Resources for Development). 2002. Available at <http://www.rcmrd.org/AboutUS/AboutUSIndex.htm>. Accessed June 25, 2002.

4

Facilitating the Use of Geographic Data: Spatial Data and Telecommunications Infrastructures

INTRODUCTION

Two infrastructures in particular facilitate the use of geographic data: spatial data and telecommunications infrastructures. This chapter begins by describing and illustrating national and global spatial data infrastructures.[1] To realize the full potential of a spatial data infrastructure (SDI) requires a telecommunications infrastructure that eases access, use, and sharing of geographic data and information. The remainder of the chapter reviews current status and future trends in telecommunications infrastructure in Africa. Within the context of telecommunications lie fixed-line and wireless telephone services, computer infrastructure, Internet infrastructure, and media tools for dissemination of geographic information. Progress in the use of geographic information for decision-making will be enhanced when efficient spatial data and telecommunications infrastructures coexist.

SPATIAL DATA INFRASTRUCTURES

An SDI is an "umbrella of policies, standards, and procedures under which organizations and technologies interact to foster more efficient use, management and production of [geographic] data" (FGDC, 1996). The following quotations emphasize the importance of SDIs in Africa:

> Building infrastructure for geo[graphic] information use is becoming as important to African countries as the building of roads, telecommunications networks, and the provision of other basic services.... The rationale for investing in information infrastructure is analogous to that for physical infrastructure: the provision of many other services is contingent upon their existence. The cost-effective development of [a spatial data infrastructure] requires the coordinated harnessing of resources and expertise residing in various government agencies, the private sector, universities, non-governmental organizations, and regional and international bodies (EIS-Africa, 2002).

> African countries are going through a familiar phase that many countries have gone through in their GIS development whereby different sectors engage in GIS activities without coordination. It is not uncommon to find different agencies collecting the same data at the same or different times (ECA, 2001).

A variety of geographic data is produced, used, maintained, and shared in application areas that include transportation, environment, natural resources, agriculture, health, emergency services, and telecommunications (Lachman et al., 2001). Data are collected and archived in varied formats. These data are primarily paper maps or in digital form and can be analyzed in GISs. If the data are to be used effectively for decision-making, they must be well managed. This goal can be accomplished by producing, organizing, storing, and distributing data cooperatively using SDI concepts (NRC, 1993).

An SDI promotes data access, use, and sharing to improve the application of geographic information by decision-makers at all scales (FGDC, 2002). A National Spatial Data Infrastructure (NSDI) is a framework that is consistent for an entire country (Table 4-1) and a Global Spatial Data Infrastructure (GSDI) is a partnership of many NSDIs.

For informed decision-making and effective governance countries require knowledge about their physical and social geography (Groot, 2001). SDIs provide a framework that facilitates these actions and make it possible to use data many times for many applications. Thus, countries can benefit economically from SDIs. For example, the government of Australia is developing the Australian Spatial Data Infrastructure to underpin the planning and management of land use, infrastructure, mining, agriculture, forestry, environment,

[1]A spatial data infrastructure consists of the technological, organizational, and management requirements that constitute the framework for a functional spatial data system.

TABLE 4-1 Countries with National Spatial Data Infrastructures

Argentina	Australia	Canada	Colombia	Cyprus	Finland
France	Germany	Ghana	Greece	Hungary	India
Indonesia	Japan	Kiribati	Macau	Malaysia	Mexico
Netherlands	New Zealand	Norway	Pakistan	Poland	Portugal
Russian Federation	South Africa	Sweden	United Kingdom	United States	

SOURCE: Longley et al. (2001); Moeller (2001).

defense, and emergency services across the country. Price Waterhouse undertook a study in 1995 to evaluate the economic benefits of Australia's investment in an SDI. The study concluded that "for every dollar invested in producing spatial data, $4 of benefit was generated in the economy. In 1989-1994 these benefits were in the order of $4.5 billion distributed across the broad spectrum of economic activities" (Nairn, 1999).

National Spatial Data Infrastructures

The concept of an NSDI is recent and acceptance by developed countries is far from complete. Developing countries can learn from the mistakes of those who have gone before and potentially bypass similar problems toward an efficient SDI (Taylor, 1997).

In the United States in 1990 the Federal Geographic Data Committee (FGDC) was created to develop a strategy for an NSDI. The goal is to have "current and accurate geospatial data that is readily available (locally, nationally, and globally) that can contribute to economic growth, environmental quality, stability, and social progress" (FGDC, 2002). The FGDC works to

(1) reduce duplication of effort by government agencies in data collection;
(2) improve quality and reduce costs related to geographic data;
(3) make geographic data more accessible to the public;
(4) increase the benefits of using available data; and
(5) establish key partnerships with states, counties, cities, tribal nations, academia, and the private sector.

The U.S. NSDI gained prominence in 1994 following the publication of U.S. presidential Executive Order 12906, which emphasized the need to coordinate the acquisition, access, and sharing of geographic data in the federal government. It elevated the NSDI from a technical subject to an essential component of social and economic development (Groot, 2001). The order indicates that all efforts described in the NSDI are to be carried out through partnerships among federal, state, and local government agencies and the public, private, and academic sectors.

Two countries in sub-Saharan Africa have formal NSDIs: Ghana (National Framework of Geographic Information Management) and South Africa (National Spatial Information Framework) (EIS-Africa, 2002). In addition, eight countries (Benin, Botswana, Ethiopia, Mali, Senegal, Swaziland, Zambia, and Zimbabwe) currently have SDI-building initiatives. Several programs also have SDI-building components, including the SADC Food Security Programme (Zimbabwe), Regional Tsetse and Trypanosomiasis Control Programme (Malawi, Mozambique, Zambia, Zimbabwe), Biomass Programme of Uganda, Kenya's Wetlands Conservation and Training Programme, the Miombo Network in southern Africa (Chapter 7), and the Community Based Natural Resources Management initiative in Namibia (Chapter 7). The barriers that hamper other countries from adopting a formal SDI include a lack of prominent SDI champions in influential positions and declines in funding for SDI-related projects.

Components of a Spatial Data Infrastructure

An SDI comprises standards, framework foundation data, framework thematic and other geographic data, metadata, clearinghouses, and partnerships.

Data Standards

Spatial data infrastructures cannot function without standards. Standards are specifications and documented practices applied to spatial data formats, data compression and decompression formats, data transmission formats, metadata formats, and computer interfaces that allow people to easily interact within the system.

A lack of standards impedes spatial data collection, distribution, and processing. Consequently, countries such as Ghana and South Africa have begun to develop standards for their spatial data. Organizations such as EIS-Africa (Box 4-1) are working to develop generic information technology-based standards (e.g., description, query language, syntax), application-independent standards (e.g., geometry, topology, metadata), and national standards tailored to specific applications (e.g., base mapping, cadastral mapping, transportation planning, and urban planning) (EIS-Africa, 2001).

Over the last 10 years several international efforts in standards development have been initiated. For example, in 1992 the European Committee for Standardization began promoting voluntary technical harmonization in Europe and

BOX 4-1
Environmental Information Systems-Africa

Focus: EIS-Africa facilitates the strategic development and use of geographic information in environmental management and sustainable development in Africa. EIS-Africa develops common principles and practices for EIS development and application, and facilitates the coordination of national and international EIS programs. Other parts of EIS-Africa's framework are documenting and sharing best practices, building on existing activities, and partnering with other organizations.

Sponsors and partners: EIS-Africa is a non-profit, pan-African organization of geo-information practitioners and institutions. Membership is open to all sectors: There are 24 member countries, mostly in sub-Saharan Africa, and over 2,000 individual and institutional members. EIS-Africa is a new organization that has evolved from the ten-year old EIS program. The original program was funded and supported by several donors and international development agencies, including the World Bank, the government of Norway, the U.S. Agency for International Development, the German Agency for Technical Cooperation, and several U.N. agencies including the U.N. Development Program, and the U.N. Environment Program (UNEP).

Key results to date: The EIS-Africa network consists of information managers, decision-makers and other professionals, as well as institutions in sub-Saharan Africa that produce or use environmental information for a variety of purposes. EIS-Africa serves as a pool of expertise, technical resources, and a knowledge base for assisting African governments and civil society to meet their priority needs for information on the environment, natural resources, and sustainable development. Numerous workshops and conferences, including the Africa-GIS series (which began in 1993), have been held to exchange information, build partnerships, and develop a network of relationships throughout Africa. Additionally, EIS-Africa has authored or co-authored publications on GIS technology applications in Africa (e.g., Bassolé et al., 2001; EIS-Africa, 2001, 2002).

Lessons learned: The term "environment" in EIS-Africa has become a limiting factor for the organization as it seeks to broaden its network to economic development, agriculture, transportation, telecommunications, and other areas of public policy. However, the organization is trying to address such issues (EIS-Africa, 2001).

The organization also has difficulties convincing governments of the value of SDI within their own borders. It is even more difficult to persuade them of the value of cross-border data sharing and cooperation on common data policy issues. Often the bureaucratic interests are defined narrowly and provincially, and the notion of regional cooperation appears to be a low priority. To address such issues, EIS-Africa sees the future of SDI in Africa in terms of building a network of field practitioners who understand the value of cooperative activity through their own experiences.

throughout the world. Most European countries support this effort. In 1994 the International Standards Organization (ISO) undertook a similar initiative (EIS-Africa, 2001). Both the European Committee for Standardization and the ISO reached an agreement in 2000 to ensure a joint international standard. Additionally, the Digital Geographic Information Exchange Standards were developed to support the efficient exchange of digital information among North Atlantic Treaty Organization nations. In 2000 the European Joint Steering Group on Spatial Standardization and Related Interoperability was created to insert spatial technology into mainstream information technology and to develop standards.

Organizations in the United States, such as the Open GIS Consortium (OGC), are developing geographic information processing interface software and data standards that operate on almost all computer platforms (in other words, making them "interoperable") (Box 4-2). As a result people can use different types of GIS software and their associated data on a variety of devices (e.g., mainframe computer, personal computer, or on a hand-held personal digital assistant). Furthermore, a GIS user can process data from dispersed sources.

The OGC has an interoperability program composed of a series of initiatives to accelerate the development and acceptance of OGC specifications (M. Reichardt, OGC, personal communication, 2002). Among these initiatives are collaborative efforts called "pilot projects" that test and implement OGC specifications (Box 4-3). Another initiative is a joint project of the European Commission's Joint Research Centre and the U.S. Geological Survey (USGS). The purpose of this project is to remove technical obstacles to sharing earth observation data. It is a direct result of the 1997 U.S. and European Science and Technology Agreement Concerning Earth Observation Technology Development and Application Research and a subsequent 1999 cooperative arrangement between the European Commission's Joint Research Centre and the USGS (M. Reichardt, OGC, personal communication, 2002).

Framework Foundation Data

Framework foundation data usually consist of three spatial data layers: (1) geodetic control, (2) digital elevation and bathymetry, and (3) digital ortho-imagery (NRC, 1995, 2001). Because people are central to Agenda 21 issues it is appropriate for human population distribution to be a fourth foundation layer (Figure 4-1).

Geodetic Control

Geodetic control is the common reference system for establishing the coordinate position (i.e., latitude, longitude, elevation) of geographic data throughout an SDI. It ties all

BOX 4-2
The Open GIS Consortium

The Open GIS Consortium (OGC) is an international industry consortium of more than 220 companies, government agencies, and universities participating in a consensus process to develop publicly available geographic information processing specifications (OGC, 2002). The OGC sets standards so that the commercially available geographic information processing software and data produced by them are interoperable. The OGC also works closely with the ISO to develop standards. The OGC is developing open, common graphical user interfaces to communicate between software system components. OGC interfaces provide access to both information and functionality. The OGC also works to develop software approaches that address inconsistent data dictionaries and metadata schemas (OGC, 2002).

geographic features and thematic infrastructure to a common horizontal and vertical coordinate system. This system allows users to locate objects (e.g., bridges, markets, mosques) and linear features and networks (e.g., political boundaries, roads, or rivers).[2]

Digital Elevation and Bathymetry

Elevation is the height above a point of reference such as Earth's surface above or below a user-specified datum (or zero line). Invariably, raw elevation data are converted into more visually understandable displays, including contour maps and shaded relief maps. Land-surface elevation is generally measured relative to mean sea level or some other datum. The elevation of terrain below a water body is referred to as bathymetric data (NSDI, 1997).[3]

[2]Each geodetic control point in a spatial data infrastructure includes name, identification code, latitude and longitude, orthometric height, ellipsoid height, and metadata. The metadata for each geodetic control station contains descriptive data, positional accuracy, and condition information.

[3]For land surfaces, the framework generally uses an elevation matrix. For depths, the framework consists of soundings and a gridded bottom model (NSDI, 1997).

BOX 4-3
Open GIS Consortium Pilot Project in Africa

The Geospatial Information for Sustainable Development Initial Capability Pilot (ICP) is the first of a series of projects to make geographic information more accessible and useful to decision-makers working on sustainable development challenges. The goal is not to create yet another network of data sources for Africa; rather, it is to create a framework of existing networks that leverage the work already accomplished by making it easier and quicker to discover, combine, and exploit existing data. ICP offers a path to this vision. It will deliver a limited but operational framework of services to illustrate how interoperable applications can improve information sharing and application in Africa. This initiative emphasizes the value of public-private partnerships and international and industry standards to make geographic information and interoperable technologies more accessible to researchers and practitioners. Often the products from different vendors meet similar requirements but fail to mesh because of vendor-proprietary formats and different processing approaches.

The ICP is funded by the U.S. Department of State and the USAID, with supporting sponsorship from Natural Resources Canada and the FGDC (on behalf of the GSDI). The ICP's implementation partners include technology companies:

- Advanced Technology Solutions (United States),
- Cubewerx (Canada),
- ESRI,
- Federal Geographic Data Committee,
- Intergraph,
- Ionic Software (Belgium), and
- Laser Scan (UK)

and organizations in Africa including:

- AGRHYMET,
- EIS-Africa,
- ICRAF (International Center for Research in Agroforestry),
- ICRISAT (International Crops Research Institute for Semi-Arid Tropics),
- CSE, Senegal (with SISEI)
- OSS
- Regional Remote Sensing Unit, Harare Zimbabwe, and
- University of Dar Es Salaam (Kenya-Tanzania Coastal Zone Management project).

- see OGC (2002) for additional partners

The intended result is a limited but operational framework of interoperable Web-based and stand-alone applications and servers operating as a single network. This will be facilitated by commercial and noncommercial software that uses OpenGIS specifications, which should greatly simplify the process of data and application sharing.

SOURCE: M. Reichardt, OGC, personal communication, 2002.

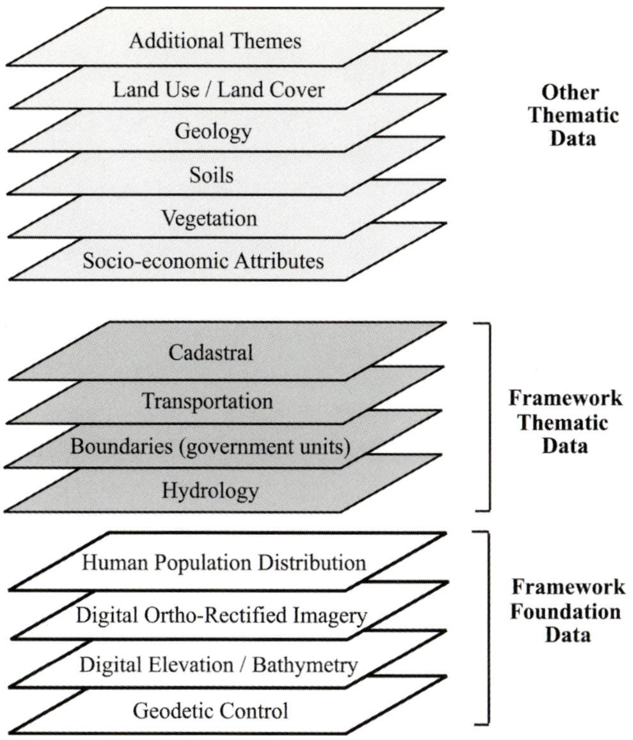

FIGURE 4-1 (Read from the bottom to the top.) A spatial data infrastructure (SDI) typically consists of framework foundation data such as geodetic control, digital elevation and bathymetry, and ortho-imagery. Because of the central nature of people in sustainable development, data on human population distribution are equally important. Where possible, SDIs also contain essential framework thematic data layers, including hydrology, political and other boundaries, transportation resources, and cadastral information. Other thematic information such as socioeconomic data, vegetation, soils, geology, and land cover may be included in the infrastructure (adapted from FGDC, 2002).

Digital Ortho-imagery

An ortho-image is a specially processed image prepared from an aerial photograph or remotely sensed image that has the metric qualities of a traditional line map with the detail of an aerial image.[4] Because ortho-images are geographically referenced, they are useful in their own right or as a backdrop upon which other information can be overlaid (e.g., drainage or road networks, utilities, or government boundaries). They also can be used as a reference base map to which other maps or images can be linked to detect changes in the landscape.

[4]During processing of the source data, an ortho-image is adjusted (rectified) to a standard map projection and datum. Geometric errors caused by topography and other anomalies are removed from the dataset during processing (Thrower and Jensen, 1976).

Human Population Distribution

Human population distribution refers to the location of people on Earth's surface. People are both influenced by and have an impact on ecosystems in which they live, and are therefore central to Agenda 21 issues. Information on the geographic distribution of the human population and their attributes are equally as important as other SDI framework foundation data. In the current worldwide development arena, such key issues as good governance, anti-poverty strategies, and the need to promote economic growth with social equity all require population distribution and other demographic data at the local scale. Geographically referenced, standardized census data that can be linked to other layers of geographic data are required to meet national development needs. Progress toward Agenda 21 goals is impeded by the lack of reliable data on human population distribution.

Framework Thematic Data

There is general consensus in the geographic information community that four of the most important framework thematic datasets are Hydrology, Boundaries, Transportation, and Cadastral data (NRC, 1995, 2001; NSDI, 1997).

Hydrology

There are three categories of hydrologic features: (1) surface water features (e.g., oceans, seas, lakes, reservoirs, and ponds), (2) linear features (e.g., shorelines, rivers, canals, and perennial and intermittent streams), and (3) point features (e.g., wells). A complete hydrologic dataset requires information about how the hydrologic network is connected and the direction in which water flows.

Boundaries

Boundaries range from the political borders of countries to administrative units to communal and individual holdings. Without accurate boundary information it is difficult to monitor an activity with a given legal jurisdiction or allocate resources fairly to people within a specific administrative district. Box 4-4 describes the importance of boundary information in Ghana.

Transportation

Transportation networks include roads, railways, waterways, and pipelines. Even in major cities of Africa they are inadequately mapped for such basic functions as delivery and collection services (ECA, 2001).

Cadastral Data

Cadastral data refer to the geographic extent of past, current, and future rights and interests of private and commercial property (FGDC, 2002). A cadastre is a map accompanied by a register showing the ownership or possession of individual units of land. It facilitates efficient land adminis-

BOX 4-4
Developing Jurisdictional Boundaries Information in Ghana

In Ghana, boundary maps (i.e., regional, district, census, electoral, and related maps) contain inaccuracies in the location of boundaries. The problem has confronted researchers, political and security administrators, and development planners for many years. Participants at the National Framework for Geospatial Information Management Steering committee consultations at Dodowa in June 2001 recommended developing a collaborative project to address the problem.

The Survey of Ghana recognizes that boundaries are an integral part of the development of a national spatial data framework for the collection, processing, and use of geographic information. Nationally accepted, accurate, and reliable district and sub-district boundary maps are needed as the basis for data collection, processing, and presentation. Presently there are as many versions of the maps as there are producers and users of information, and it is never clear which version is used in national and private documents.

SOURCE: <http://www.epa.gov.gh/Nafgim/Ongoing_Acts.htm>.

tration and expedites land-market transactions. The spatial data necessary to describe the geographic extent of property rights and interests includes surveys, legal description reference systems, and parcel-by-parcel surveys and descriptions.

Metadata

Metadata (defined as data about data) describe existing data holdings, making it possible for people outside an agency or organization to access, search, and use geographic data. Metadata help people know that data exist; how, when, or why the data were produced; and where the data reside. For example, when making important land-management decisions, it is helpful to know where to find the needed geographic data (Table 1-1).

A U.S. standardized metadata format, Content Standards for Digital Geospatial Metadata, was approved by the FGDC in 1998. An international metadata standard became available in 2002, and the National Spatial Information Framework of South Africa is compiling a metadata profile based on these standards.

Clearinghouses

Clearinghouses (or catalogs) allow government agencies, non-profit organizations, and commercial participants worldwide to make their geographic data searchable and accessible through the Internet. A clearinghouse carries the potential to reduce duplication of effort in collecting and distributing digital spatial (and non-spatial) data.

Clearinghouses serve three principal functions. First, they provide an Internet-based graphical user interface helping users search for (or "discover") geographic data with standardized metadata. Second, they allow access to raw digital data and allow standardized access to geographic data through file-transfer-protocol[5] directories and online data streaming services. Third, users may access data and process geographic information to make maps over the Internet.

In developed countries national data clearinghouses are now commonplace. For example, a shared data discovery infrastructure was set up across Europe in 1998 (the European Spatial Metadata Infrastructure [ESMI] project). The ESMI project involves mapping organizations of most European countries and provides a common research and development framework to permit discovery of geographic data. Clearinghouse efforts are also underway in the developing world (Box 4-5).

Partnerships

Partnerships bind spatial data infrastructures together at all scales. Groups and government agencies enter into partnerships to freely share geographic data. These partnerships form a basis for many spatial data infrastructures (Table 4-1). The partnership concept is extended worldwide through the GSDI.

Global Spatial Data Infrastructure

The GSDI is an emerging network of public and private national, regional, and international organizations. It involves the development and open sharing of global to local data through a network of clearinghouses. The GSDI advocates the adoption of appropriate standards and practices for sharing data. Currently the secretariat is with the U.S. FGDC (see Appendix C).

The GSDI includes such regional groupings as the European Umbrella Organization for Geographic Information, Permanent Committee on GIS Infrastructure for Asia and the Pacific, Permanent Committee on Spatial Data Infrastructure for the Americas, and the Permanent Committee on GIS Infrastructure for Africa. Such groupings enable participants to benefit from shared experiences and lessons learned in the application of geographic data.

An implementation guide, *Developing Spatial Data In-*

[5]A standard protocol and an application that permits files to be copied from one computer to another, regardless of file format or operating system.

BOX 4-5
Clearinghouse Activities in Africa

Under the auspices of the U.N. Environment Programme (UNEP) in Nairobi, Kenya, a global directory of environmental data is being prepared using UNEP-sponsored metadata and software. Through international collaboration, conversion of the software is underway to make it compatible with FGDC metadata standards. The UNEP metadata holdings will be made available through a search service on the Internet.

In South Africa the National Spatial Information Framework project coordinates the development of metadata and an online discovery system. This system uses FGDC metadata standards and related software and supports discovery of geographic data in South Africa and in the greater southern Africa region.

The CEOS Information Locator System, built by the International Committee on Earth Observation Satellites, enables users in developing countries to gain access to Earth observation information. Users can also enter, administer, and share their own data and information. Metadata host sites are maintained in China, Norway, Kenya, Japan, Italy, Germany, and Australia, providing regional access to a common and synchronized international collection of metadata. A majority of the sites are using a common method for searching for data.

SOURCE: FGDC (2001).

frastructure: The SDI Cookbook, was released by the GSDI Secretariat during the GSDI-5 conference held in Cartagena, Colombia, in 2001 (GSDI, 2002). The document introduces local, regional, national, and multinational organizations to the common concepts and issues, policies, standards, and recommendations for implementing a globally compatible SDI. The guide provides details on how to establish compatible systems within and between organizations to facilitate information discovery, applications support, and exchange. It also identifies, describes, and references relevant standards and specifications from the ISO, Open GIS Consortium, and the Internet to assemble a coherent vision for their integration (GSDI, 2002).

The GSDI is economical because it enables worldwide cost sharing of data creation and maintenance: Those who produce the data maintain and update it. As a result of the potential benefits of both the GSDI and SDI concepts, **developing countries should consider using a standardized SDI that is compatible with the emerging GSDI. Data derived from international development programs (for example, those of USAID) should conform to the standards recommended by the GSDI.** In this way data collected by these programs should be rendered more useful.

TELECOMMUNICATIONS INFRASTRUCTURE IN AFRICA

SDIs depend on efficient telecommunications infrastructures (Figure 4-2). In particular the accumulation of data and information without the ability to access, process, and disseminate serves little purpose. In general, telecommunications infrastructures in the developing world are inadequate to support SDIs. Telecommunications technologies that meet the needs for geographic data transfer and information dissemination are vital to close the digital divide between developed and developing countries, between urban and rural areas, and between the rich and the poor (InfoDev, 2001). As a result of this divide the potential impact of geographic information and associated technology on development challenges in Africa are largely unrealized.

Overview

Telecommunications infrastructures in Africa have generally improved during the 1990s (Table 4-2). For example, use of satellite television, the Internet, personal computers, and fixed and cellular telephones is increasing across the continent (Taylor, 1997; ECA, 1999a). Of the roughly 800 million people in Africa (Jensen, 2002)

- 1 person in 13 has a television (62 million),
- 1 in 40 has a fixed telephone line (20 million),
- 1 in 35 has a wireless telephone (24 million),
- 1 in 130 has a personal computer (6 million),
- 1 in 160 uses the Internet (5 million), and
- 1 in 400 has pay television (2 million).

As elsewhere, telecommunications in African countries correlate with the settlement hierarchy. For example, in Tanzania the adoption of telecommunications is greatest in Dar es Salaam, the economic and social nerve center of the country. In such provincial cities as Arusha, Iringa, Tanga, and Mwanza the adoption of these technologies is lower. In rural areas of Africa where 7 out of 10 people live, levels of adoption are lowest (Jensen, 1999) and much of rural Africa has been isolated from international networks. Africa's diversity is an important characteristic that represents both a strength and a challenge to developing telecommunications infrastructures (Box 4-6). Fortunately adoption of telecommunications technologies will increase in the future because of

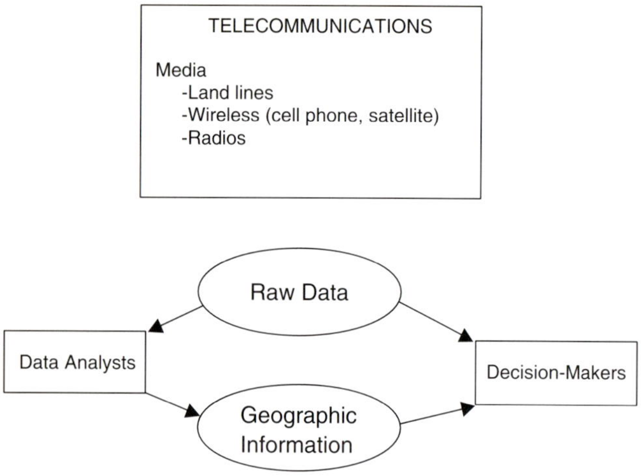

FIGURE 4-2 Ways that geographic data and information may be transferred from NSDI or GSDI clearinghouses to African decision-makers.

TABLE 4-2 Recent Telecommunication Trends in Sub-Saharan Africa

	1996	1999	2000
Fixed line and mobile telephones (per 1,000 people)	13.4	24.3	31.6
Personal computers (per 1,000 people)	Not available	8.2	9.2
Internet Users	648,000	2.4 million	3.7 million

SOURCE: Development Gateway (2002).

declining unit costs, policy changes and liberalization of markets (ECA, 2001; InfoDev, 2001).

There are growing numbers of kiosks, cyber cafés, and telecenters that offer information and communications technology access and services to the public for a fee. At these venues people can access geographic data and information in villages and participate in technology training to address local development needs. Available technologies include telephones, fax machines, computers, photocopiers, email, and the Internet. In Ghana, for example, wireless pay-phone kiosks provide an inexpensive way to broaden rural access to the networked world (McConnell International, 2000). Regulatory policies that encourage increased telecommunications access and new administrative arrangements allow entrepreneurs to operate profitable enterprises.

Continued improvements in telecommunications infrastructures could enhance the "e-readiness" of African countries to support e-government and e-business. In *The Global Information Technology Report 2001-2002: Readiness for the Networked World* (World Economic Forum, 2002), Harvard University ranked countries for e-readiness[6] based on their capacity to take advantage of information and communication technology networks. Four African countries were among the 75 that were ranked: South Africa, 40th; Egypt, 60th; Zimbabwe, 70th; and Nigeria, 75th. With policy reform and regulatory change African countries could become globally competitive in e-commerce, especially in services (ECA, 2001).

Other organizations are developing and improving telecommunications infrastructure in Africa. These include the African Information Society, African Development Forum, African Telecommunications Union, African Connection, ECA, Economic Commission for West African States, and Common Market for Eastern and Southern African (Box 4-7). International organizations that improve telecommunications infrastructure include the International Telecommunications Union, UNESCO, and the World Bank. In addition, InfoDev funds pilot projects and policy, networking, and human resource development activities that promote the dis-

[6] E-readiness describes the extent to which a country's business environment is conducive to Internet-based commercial opportunities. It is a concept that spans a wide range of factors, from the sophistication of the telecommunications infrastructure to the security of credit-card transactions and the literacy of the population (<http://www.ebusinessforum.com>).

BOX 4-6
Africa's Diversity

Many Africans identify themselves as members of an ethnic group (one or more depending on parentage), name the language they speak as a member of that group, and identify the place (village, camp, town, city) where they live as significant. Commonly the ethnic group name and the language they speak are the same. Importantly, most Africans speak two or more languages—the home language, the lingua franca, and the language of the nation. In addition, many speak English in Anglophone Africa, French in Francophone Africa, or Arabic in North Africa.

The number of ethnic groups in Africa is difficult to know precisely because estimates vary. Armstrong (1963) estimated between one thousand and two thousand different languages. For purposes of this report, the estimate of 1,000 ethnic groups is a reasonable minimum. In addition to these groups, such European settlers as the English in east and South Africa, the Boers in South Africa, and the Asians in eastern and South Africa are significant in their national populations and influence on linguistic diversity.

BOX 4-7
Regional Economic Organizations and Telecommunications

U.N. Economic Commission for Africa (ECA)

In 1996 the ECA adopted an initiative to create an information infrastructure for Africa through access to a global infrastructure. The goal is to provide information for development and to build national communications capacity in member states. The ECA emphasizes the importance of the involvement of government, non-government, and private-sector institutions as well as the science and technology communities. The plan also stresses the need for policy and legislative reform at the national level to promote effective decision-making (ECA, 1999b).

Economic Commission for West African States (ECOWAS)

The Economic Commission for West African States spearheaded the INTELCOM I programme to connect the capital cities of West Africa by microwave links. Subsequently, ECOWAS recommended that priority be given to completing the INTELCOM II programme to facilitate telecommunications development through improved satellite or fiber-optics links. The goal of this program is to provide regional infrastructures to stimulate trade and investments, in part through involvement of the private sector. The program strives to connect existing networks in the areas of transportation, communications, and energy (ECOWAS, 2002).

Common Market for Eastern and Southern Africa (COMESA)

The Common Market for Eastern and Southern Africa is promoting the establishment of a regional telecommunications network to increase trade among the member states. The network includes a mixture of fiber-optic, microwave, and satellite connectivity that will facilitate the transfer of data and television programming. Where possible, the network follows the layout of existing infrastructure. The network is administered by COMTEL Communications, Ltd. and private-sector investors will have the majority equity shareholding (COMESA, 2002).

semination of information and communication technologies. Finally, Volunteers in Technical Assistance (VITA)[7] and Wavix, Inc.,[8] are launching the VITA Connect Network to provide information and low-cost, remote-area connectivity for development and humanitarian organizations. The activity will rely on its Low Earth Orbiting Satellite—the first satellite operated by a non-governmental organization focusing on development (VITA, 2002).

The U.S. government also is enhancing African telecommunications infrastructure. For example, the Leland Initiative, a five-year, $15 million activity, promotes Internet connectivity and its sustainability in 21 African countries. Currently, regulatory reform is the focus of the initiative (USAID, 2002). In another example, the Radio and Internet for the Communication of Hydro-meteorological and Climate Related Information Project (RANET) provides information access and support to rural communities on topics such as weather, agriculture, education, and health (RANET, 2002). NOAA, the University of Oklahoma, and USAID provide technical and financial support to RANET.

The U.S. government (e.g., USAID and NOAA) should continue to assist African countries in improving telecommunications infrastructure so that large computer files containing geographic data can be readily distributed within national and global spatial data infrastructures.

Fixed-line and Wireless Telephone Service

In 2001 total telephone use (fixed-line plus wireless telephones) per 100 people was 121.1 in developed countries and 19.8 in developing countries (ITU, 2002b). In Africa the countries of the Sahel and central Africa, such as Niger and Democratic Republic of Congo, have less than two telephone lines for every 1,000 people. Northern Africa and South Africa have a teledensity[9] around 35 per 1,000 people, and West and East African coastal countries have teledensities between 2.5 and 10 per 1,000 people (ECA, 1999a). When the Commission for Worldwide Telecommunications Development published the Missing Link Report in 1984, there were more telephones in Tokyo than in all of Africa. Today there are twice as many telephones in Africa than in Tokyo (ITU, 2002a).

Although charges for local and international phone calls are declining, they remain high in comparison with rates in Europe and North America (ITU, 2001). For most Africans, making a telephone call is an unaffordable luxury. Telephone subscription costs range from one-sixth to nearly one-half of per capita gross domestic product (ITU, 2001). Moreover, the quality of telephone service is variable (ITU, 2001), and the average waiting time for the installation of a telephone is 3.5 years in sub-Saharan Africa (ITU, 2001).

Fortunately, wireless technology is expanding telephone networks in Africa. In 2001 most countries in sub-Saharan

[7]An international development organization in the U.S. state of Virginia.

[8]A private firm that developed a two-way satellite communication system specifically to support remote data collection and e-mail services.

[9]Teledensity is the number of telephone connections per unit population (e.g., per 1,000 people), typically measured by the number of fixed lines per unit population, number of wireless phones per unit population, or both.

FIGURE 4-3 Countries with more wireless (mobile) than fixed telephone lines in 2001. SOURCE: ITU (2002a).

Africa had more wireless (mobile) than fixed-wire subscribers (Figure 4-3). The success of this technology relates to the lower costs of providing wireless coverage rather than installing fixed copper lines, and to liberalization of telecommunications regulations and the resulting competition in wireless services.

Computer Infrastructure

Worldwide the cost of computers is decreasing and the number of computers is increasing (Taylor, 1997). In northern Africa and South Africa there are more computers per capita than in the rest of Africa (Table 4-3). For example, South Africa has roughly 6 computers per 1,000 people, whereas other sub-Saharan countries combined have less than 1 computer per 1,000 people. By comparison, the United States had 459 computers per 1,000 people in 1998 (World Bank, 2001, Table 19).

Typically, computers are imported into African countries. Consequently, prices are high when compared to per capita gross domestic product. The high cost of computers limits the rate of Internet diffusion and access to geographic data clearinghouses. Most personal and mainframe computers are in South Africa, and most of them are confined to ministries of finance (for government payroll), large parastatals, telecommunications operators, banks, and insurance companies (Jensen, 1999).

The high cost of new computers is offset to some extent by organizations that transfer refurbished computers to Africa without charge or for a nominal fee. For example, Computer Aid International has distributed a total of 1,800 used computers to Ghana, Mozambique, South Africa, and Zimbabwe. Additionally, open-source software[10] has become increasingly important because development and maintenance costs are incurred by communities of volunteers rather than by users, and the software is available either without charge or for the cost of disks, manuals, and technical support. Additionally, by using the Internet Africans can use computers in Europe and North America.

A range of challenges face computer users in Africa. Poor computer maintenance, insufficient skills to diagnose system problems, and underuse of existing computers are commonplace (ECA, 1999a). These problems are the result of insufficient financial resources, limited training in computer maintenance and software trouble-shooting, and lack of technical support by computer manufacturers. In sub-Saharan Africa environmental conditions (such as dust, heat, and heavy rains) and irregular or nonexistent electricity supplies also hamper computer use and disrupt Internet connections. Power outages that last many hours are a regular occurrence, even in capital cities (ECA, 1999a), and poor transportation networks hinder distribution of physical goods.

TABLE 4-3 Personal Computers in Africa

	Personal Computers (1000s)	Personal Computers per 100 people
Northern Africa	2,169	1.64
South Africa	2,700	6.18
Other Sub-Saharan	2,455	0.47
All Africa	7,324	1.05

SOURCE: ITU (2001).

Internet Infrastructure

In November 1991 South Africa established the first Internet connection in Africa and continues to lead in

[10]Software for which the original (source) code is freely available.

TABLE 4-4 Internet in Africa in 2001

Region	Dial-Up Internet Subscribers	International Outgoing Bandwidth Kbps	Number of Internet Service Providers	Population millions	Cities with Points of Penetration	Dial-Up Subscribers per Million People
Central	10,100	3,072	17	71.48	12	141
East	120,600	24,894	80	216.42	28	557
North	279,000	294,096	360	139.25	36	2,004
South[a]	80,350	33,044	35	69.42	38	1,157
South Africa	750,000	350,000	80	44.31	100	16,926
West	122,725	68,072	77	228.78	34	536
Africa	1,362,775	773,178	649	769.66	248	1,771

[a]Exclusive of South Africa.

SOURCE :National Intelligence Council (2001).

Internet-related developments (Table 4-4). Eritrea, the last country to be connected, linked up in November 2000 (ITU, 2001). Currently there are about 649 Internet service providers in Africa and the number will continue to increase as Internet markets open.

The number of Internet users in Africa is difficult to estimate because there may be several users per subscription and because of the popularity of public facilities (e.g., kiosks, cyber cafés, and telecenters). In 2001 Africa had an estimated 4.4 million Internet users. More than half of these users lived in South Africa, and one-sixth of them lived in the Maghreb countries. An estimated 1.3 million users were in sub-Saharan countries (excluding South Africa) (ITU, 2001). Combined, these numbers account for 0.15 percent of total global Internet connectivity (ITU, 2001).

International bandwidth[11] for African countries is low (Figure 4-4). For example, the 400,000 people living in Luxembourg share more international Internet bandwidth than Africa's 760 million people (ITU, 2001). Until recently few countries outside South Africa had international Internet links larger than 64 kilobytes per second (equal to a common capacity of telephone modems in North America), but in 2002 nearly half the countries have links of 2 megabytes per second or more, and 20 percent have outgoing links of 5 megabytes per second or more (Jensen, 2002). Because it can be measured reliably, bandwidth can be used to indicate a country's progress toward increased access to information.

In Africa satellite-based Internet access is increasing. For example, satellite dishes provide incoming bandwidth of 64 kilobytes per second for about U.S.$30-$1,000 per month, which is cheaper than services available from local operators (Jensen, 2002). Two-way satellite-based Internet services use very small aperture terminals (VSAT)[12] to connect directly to the United States or Europe when regulations permit. The adoption of VSATs is greatest in the Democratic Republic of the Congo, Ghana, Mozambique, Nigeria, Tanzania, Uganda, and Zambia. A number of consumer-oriented VSAT services are scheduled. The target price is currently between U.S.$700-$900 for VSAT equipment providing 56 kilobytes per second outgoing capacity and 200-400 kilobytes per second incoming capacity (Jensen, 2002).

Despite increasing capacity, bandwidth continues to limit data transfer of large geographic data files across the Internet. For example, a single Landsat 7 Enhanced Thematic Mapper Plus (ETM^+) satellite image contains 390 megabytes of data. Similarly, a single Space Imaging IKONOS-2 (11 × 11 km area, 1 × 1m resolution) panchromatic satellite image is 121 megabytes.[13] Using file-compression technologies and targeting data that are most relevant to the application are two approaches that ameliorate this problem.

Cost is another factor that limits transfer of data across the Internet. In Africa the average cost for a local dial-up Internet account is about $68 a month for 20 hours (usage fees and local-call telephone time are included but not telephone line rental) (Jensen, 2002). Internet service provider subscription charges vary from $10 to $100 a month. These variations reflect different levels of market maturity, variations in tariff policies among telecommunications operators, and different regulations on private wireless data services and on access to international telecommunications bandwidth (Jensen, 2002). According to the International Telecommunications Union (2001):

> Internet access prices are generally exorbitant on the continent especially relative to the prevailing low per capita incomes.... There is a double whammy because, in addition to Internet access, dial-up users must also pay the cost of a telephone connection and telephone usage charges.

[11]Bandwidth is a measure of the amount of data that can be reliably transmitted through a channel per unit time, typically measured in bytes per second.

[12]VSATs are direct links to communications satellites, typically using antennas less than 2 meters in diameter.

[13]Seventeen respondents to the committee's inquiries in case-study regions indicated that compact disks are commonly used for data transfer, whereas only 12 of these also use the Internet.

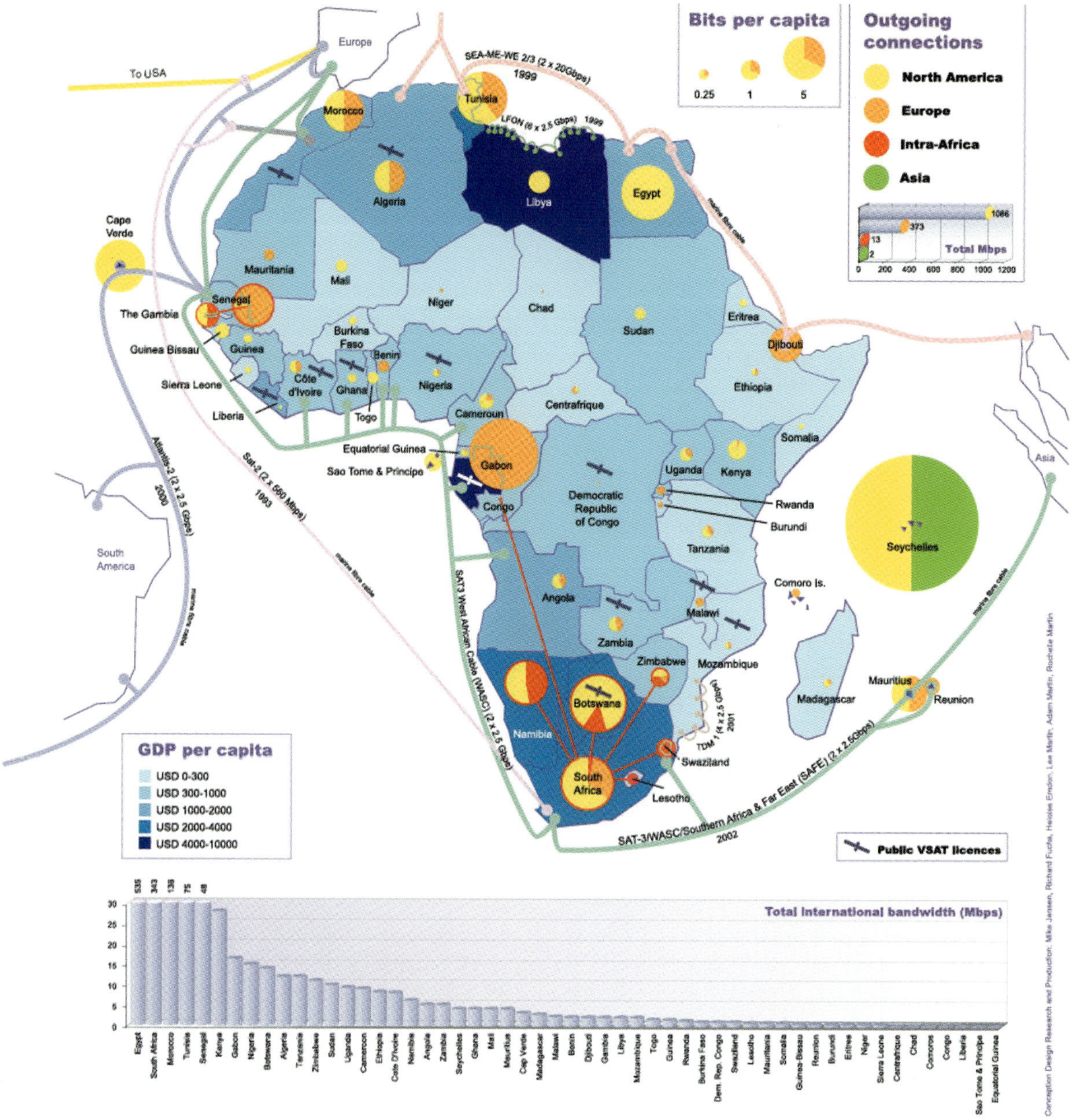

FIGURE 4-4 Internet bandwidth in African countries in March 2002. The colored circle in each country shows the international bandwidth in bits per capita (8 bits = 1 byte). SOURCE: <http://www.idrc.ca/acacia/divide/>.

Internet prices are high for three reasons. First, African countries must pay the cost of international Internet bandwidth. Second, countries do not benefit from economies of scale and do not have bargaining power to leverage cheaper prices because demand for the Internet is low. Third, landlocked countries are unconnected to the marine cables that ring the continent; therefore, Internet connectivity is limited to more expensive satellite connections.

Internet growth will continue in Africa and will be driven by technological innovations, the growth of the global Internet, and an increasingly favorable policy environment. Nonetheless, for now at least the preferred method of geographic data transfer is the compact disk.

Media

Media such as radio, television, and print are important for mass dissemination of geographic information (derived from analyses of raw data) on such topics as food security, human and livestock health, weather, environmental quality, and potential or evolving natural disasters. The audience for this information includes farmers, natural resource manag-

TABLE 4-5 WorldSpace Satellite Radio Partnership Activities

Partnering Organization	Activities	Goals
African Center for Meteorological Forecasting (<http://www.acmad.ne/>)	Provides local communities with wind-up radios and small transmitters to transmit weather data.	Provide local populations with field weather stations to collect climate information and food security data from their villages and transmit this by laptop computers to a satellite for re-transmission to a ground station.
Arid Lands Information Network (<http://www.alin.or.ke/>)	Helps local communities transmit and receive information on disease, rainfall, crops and livestock, and other data.	Improve the local economy.
SATELLIFE (<http://www.healthnet.org/>)	Provides a public health channel in four countries (Ethiopia, Kenya, Uganda, and, Zimbabwe).	Connect health practitioners to information about the treatment and prevention of such diseases as HIV/AIDS, malaria, and tuberculosis.

ers, and the general public. Radio, with the largest audience, is the primary source of information.

Media statistics should be interpreted with care because people commonly share information sources. For example, there are radio-listening clubs (Communication Initiative, 2002a), and newspaper readership is more than 10 people per paper (ECA, 1999a).

Radio

Radio is the most accessible and cost-effective means of mass communication. It overcomes barriers of distance and isolation, illiteracy, and language diversity, and is a source of information on such topics as weather, agriculture, health, local development, and education. The number of radios increased from 33 million in 1970 (93 per 1,000 people) to 158 million in 1997 (216 per 1,000 people) (UNESCO, 1999). By 1999 radio transmitter networks reached over 60 percent of sub-Saharan Africa's population (Jensen, 1999).

Radio stations fill an important role in broadcasting to rural populations. For example, Radio Douentza in lightly settled northern Mali serves an area of 15,000 square kilometers using one 250-watt transmitter (Communication Initiative, 2002b). In addition, satellite-based broadcasting is available, though expensive. With a receiver costing U.S.$225-$375, over 100 audio and visual digital channels are accessible from WorldSpace Corporation's AfriStar satellite. WorldSpace receivers provide digital audio channels and can serve as a modem for the Internet (Slifer-Mbacke et al., 2000). WorldSpace Foundation is partnering with several organizations, such as the African Center for Meteorological Forecasting, Arid Lands Information Network, and SATELLIFE to increase information dissemination in Africa (Table 4-5).

Television

Television provides a convenient and effective means of conveying geographic information. The coverage of television in Africa is similar to radio but more restricted to major urban centers. The number of television sets in Africa increased from 4.5 per 1,000 people in 1970 to 60 per 1,000 people in 1997 (UNESCO, 1999). By comparison, in1997, developed countries had 548 sets per 1,000 people.

New technologies such as digital television broadcasting systems offer a range of services, including transmission of Web pages, e-mail, and graphics. News and information can be obtained from WorldSpace's Channel Africa, a satellite-based channel that was launched last year. African Journal, a weekly one-hour call-in television program broadcast live by stations in 27 sub-Saharan African countries, focuses on Agenda 21 issues.

Print Media

Although less widespread than radio and television, print also is a valuable medium for information dissemination. Print is especially useful for distributing maps, tables, and figures. Because rural populations have limited access to this medium, most readers are from urban areas. It is estimated that newspapers have a circulation of 12 million (16 papers per 1,000 people) (UNESCO, 1999).

Convergence of Telecommunications Technologies

The convergence of telecommunications technologies is important because it increases the options for transferring and disseminating geographic data and information. For example, e-mail can be delivered by wireless phone, and voice mail can be combined with public-access telephones to disseminate information to rural areas.[14] These technologies play a similar role to radio wherein information generated from data analysis is communicated in a non-spatial format.

The Internet and other media also are linked. Radio stations, newspapers, and journals have Web sites and obtain information from the Internet. Radio stations also download audio and broadcast over the Internet. In Mozambique, the Internet is accessible over cable. In other words, there are

[14]For example, see Voxiva's pilot project in Latin America (<http://www.voxiva.net/>).

multiple ways of conveying geographic information to the public, and their reach is expanding.

SUMMARY

Spatial data infrastructures (SDIs) are a popular framework for managing geographic data and information. They exist on the national and global levels (NSDI and GSDI, respectively). African countries are at various points in development of their SDIs. SDIs and telecommunications infrastructures are intimately linked. To realize the full potential of an SDI requires a telecommunications infrastructure that facilitates access, use, and sharing of geographic data and information. Although telecommunications infrastructures are improving in Africa, they currently limit transmission of vital data and information for Agenda 21 issues. However, Africa's telecommunication links to the rest of the world likely will improve with a large number of international telecommunication infrastructure building initiatives that have been announced in the last few years (Jensen, 2002).

The next two chapters examine the rich array of geographic data sources that could be incorporated into African SDIs for effective application to environmental and developmental challenges as long as the necessary infrastructure is in place.

REFERENCES

Armstrong, R .G. 1963. Vernacular Languages and Cultures in Modern Africa. In Language in Africa, J. Spencer, ed. Cambridge, England: Cambridge University Press

Bassolé, A., J. Brunner, and D. B. Tunstall. 2001. GIS: Supporting Environmental Planning and Management in West Africa. Washington, DC: World Resources Institute.

COMESA (Common Market for Eastern and Southern Africa). 2002. COMESA Transport and Communications. COMTEL Project Brief. Available at <http://www.comesa.int/comms/comtelb.htm>. Accessed July 9, 2002.

Communication Initiative, The. 2002a. DTR Radio Listening Clubs in Zambia. Available at <http://www.comminit.com/drum_beat_141.html >. Accessed June 27, 2002.

Communication Initiative, The. 2002b. Radio Douentza: Mali. Available at <http://www.cominit.com/pdsmay15/sld-1210.html>. Accessed July 2, 2002.

Development Gateway. 2002. Available at <http://www.development gateway.org/>. Accessed May 1, 2002.

ECA (Economic Commission for Africa). 1999a. Strengthening Africa's Information Infrastructure. Report from African Development Forum. Available at <http://www.uneca.org/adf99/infrastructure.htm>. Accessed January 28, 2002.

ECA. 1999b. National Information and Communications Infrastructure in Africa. Developing National Information and Communications Infrastructure (NICI) Policies, Plans and Strategies: the "why" and "how". Available at <http://www.uneca.org/aisi/nici> Accessed July 9, 2002.

ECA. 2001. The Future Orientation of Geoinformation Activities in Africa: Synthesis. Available at <http://www.uneca.org/eca_resources/Conference_Reports_and_Other_Documents/disd/codi/docs/doc3EN.pdf>. Accessed May 15, 2002.

ECOWAS (Economic Commission for West African States). 2002. Achievements of ECOWAS: Development of Physical Infrastructure for Roads, Telecommunications, and Energy. Available at <http://www.ecowas.int/sitecedeao/english/achievements-2.htm>. Accessed July 9, 2002.

EIS-Africa. (Environmental Information System-Africa). 2001. Environmental Information Systems Development in Sub-Saharan Africa: Approaches, Lessons, and Challenges. E. Gavin and J. Gyamfi-Aidoo, eds. EIS-Africa.

EIS-Africa. 2002. Geo-information Supports Decision-making in Africa: An EIS-Africa Position Paper (draft). Available at <http://www.eis-africa.org/DOCS/EIS-AFRICAwssd_statement-draft8.doc>. Accessed June 27, 2002.

FDGC (Federal Geographic Data Committee). 1996. The National Spatial Data Infrastructure. Available at <http://www.fgdc.gov/nsdi/nsdi.html>. Accessed May 31, 2002.

FGDC. 2001. Clearinghouses in Europe and Africa. Available at <http://www.fgdc.gov/clearinghouse/ participation/eurafrica.html>. Accessed April 10, 2001.

FGDC. 2002. Overview: What the Framework Approach Involves. Available at <http://www.fgdc.gov/framework/overview.html>. Accessed April 24, 2002.

Groot, R. 2001. Reform of Government and the Future Performance of National Surveys. Computers, Environment and Urban Systems 25(4-5):367-387.

GSDI (Global Spatial Data Infrastructure). 2002. Developing Spatial Data Infrastructure: The SDI Cookbook, D. Nebert, ed. Available at <http://www.gsdi.org/pubs/cookbook/cookbook0515.pdf >. Accessed June 28, 2002.

InfoDev (Information for Development Program). 2001. Annual Report 2001. Washington, DC: World Bank.

ITU (International Telecommunication Union). 2001. African Telecommunication Indicators 2001. Geneva: ITU.

ITU. 2002a. The New Missing Link: The Digital Divide. Conference proceedings: World Telecommunication Development Conference, March 18-27. Available at: <http://www.itu.int/ITU-D/conferences/wtdc/2002/brochure/missing_link.html>. Accessed May 18, 2002.

ITU. 2002b. World Telecommunication Development Report: Reinventing Telecoms. 6th Edition. Geneva: ITU.

Jensen, M. 1999. ICT Infrastructure in Africa. Available at <http://www3.sn.apc.org/africa/intra.htm>. Accessed May 3, 2002.

Jensen, M. February 2002. The African Internet: A Status Report. Updated version. Available at <http://www3.wn.apc.org/africa/>. Accessed May 3, 2002

Lachman, B. E., A. Wong, D. Knopman, and K. Gavin. 2001. Lessons for the Global Spatial Data Infrastructure: International Case Study Analysis. Washington, DC: RAND Science and Technology Policy Institute.

Longley, P.A., M.F. Goodchild, D.J. Maguier, and D.W. Rhind. 2001. Geographic Information Systems and Science. New York: John Wiley and Sons.

McConnell International. 2000. Risk E-Business: Seizing the Opportunity of Global E-Readiness. Available at <http:/www.mcconnell international.com/ereadiness/EReadinessReport.htm>. Accessed May 13, 2002.

Moeller, J. 2001. Spatial Data Infrastructures: A Local to Global View. Available at <http:// gsdi.org/PPT/unrcc01.ppt>. Accessed May 31, 2002

Nairn, A. 1999. Commonwealth Government Initiatives to Support the Development of the Australian Spatial Data Infrastructure. Commonwealth of Australia. Available at <http://www.auslig.gov.au/corpinfo/publications/technical/docs/msasdi.pdf>. Accessed July 1, 2002.

National Intelligence Council. 2001.Workshop on Information Technology in Africa: Conference Report (unpublished).

NRC (National Research Council). 1993. Toward a Coordinated Spatial Data Infrastructure for the Nation. Washington, DC: National Academy Press.

NRC. 1995. A Data Foundation for the National Spatial Data Infrastructure. Washington, DC: National Academy Press.

NRC. 2001. National Spatial Data Infrastructure Partnership Programs: Rethinking the Focus. Washington, DC: National Academy Press.

NSDI (National Spatial Data Infrastructure). 1997. Framework: Introduction and Guide. Washington, DC: Federal Geographic Data Committee.

OGC (Open GIS Consortium). 2002. Available at <http://www.opengis.org/index.htm>. Accessed June 25, 2002.

RANET (Radio and Internet for the Communication of Hydro-meteorological and Climate Related Information Project). 2002. Via satellite: RANET Climate Information Pages. Available at <http://www.oar.noaa.gov/spotlite/archive/spot_ranet.html>. Accessed May 2, 2002.

Slifer-Mbacke, L., H. Ladd, and R. Bourke. 2000. Worldspace Foundation, SATELLIFE Create First Public Health Channel for Entire African Continent. Available at <http://www.procaare.org/newsview.php/32/>. Accessed July 11, 2002.

Taylor, D. R. F. 1997. Modern Cartography, Policy Issues and the Developing Nations. Pp 185-213 in Policy Issues in Modern Cartography, D. R. F. Taylor, ed. Oxford: Pergamon.

Thrower, N. J. and J. R. Jensen. 1976. The Orthophoto and Orthophotomap: Characteristics, Development, and Aspects of Cartographic Communication. American Cartographer 3(1):39-56.

UNESCO (U.N. Educational, Scientific, and Cultural Organization). 1999. UNESCO Statistical Yearbook, 1999. Available at <http://www.uis.unesco.org/en/stats/stats0.htm>. Accessed May 31, 2002.

USAID (U.S. Agency for International Development). 2002. Leland Initiative Website. Available at <http://www.usaid.gov/leland/>. Accessed May 2, 2002.

VITA (Volunteers in Technical Assistance). 2002. Available at <http://www.vita-connect.org>. Accessed July 11, 2002.

World Bank, 2001. World Development Report 2000/2001: Attacking poverty. Available at <http://www.worldbank.org/poverty/wdrpoverty/report/index.htm>. Accessed June 5, 2002.

World Economic Forum. 2002. The Global Information Technology Report 2001-2002: Readiness for the Networked World. G. Kirkman, P. Cornelius, J. Sachs, and K. Schwab, eds. Available at <http://www.cid.harvard.edu/cr/gitrr_030202.html>. Accessed July 1, 2002.

5

Geographic Data for Sustainable Development I: Framework Data

INTRODUCTION

This chapter examines sources of framework geographic data, their characteristics, and their adequacy for existing and potential Agenda 21 applications. Their availability and cost to users in developing countries, and the ease with which they can be updated (including the likely continuity of the data source) are considerations for the adequacy of these data.

The chapter begins by recognizing the benefits of preserving Africa's historical legacy data (e.g., maps, aerial photographs, and reports), which can serve as baseline information for measuring environmental change. The remainder of the chapter describes sources of the four framework foundation data layers (Figure 4-1).

1. Geodetic control (including geographic location and elevation);
2. Digital ortho-imagery;
3. Digital elevation and bathymetry; and
4. Human population distribution,

and the four framework thematic data layers:

1. Hydrology;
2. Government units;
3. Transportation networks; and
4. Cadastral information.

Much of the technical information on data sources in Chapter 5 is found in Annex 5.

HISTORICAL LEGACY DATA AS A BASELINE FOR DOCUMENTING CHANGE

The earliest baseline against which future change can be compared often comes from historical legacy data (Box 5-1). In many instances legacy data may be digitized, placed in a GIS, and analyzed in conjunction with more recent geographic data, such as satellite remotely-sensed data. The time scales over which change can be detected are extended through use of legacy data. A better understanding of environmental changes[1] and what could be done to ameliorate problems is gained through comparison of these data with modern records. Maps created in the 1960s and 1970s may be less accurate and offer less complete coverage than those made today, but they can provide the baseline for geodetic control, elevation, and feature recognition. Legacy maps contain place names and provide valuable insights on ethnicity and population growth. In addition, information on natural resources and socioeconomic history come from legacy data in the form of oral traditions.

In Africa legacy data come from three periods: pre-colonial, colonial, and post-colonial. Information from the pre-colonial period is mostly derived from oral traditions, and this is still disseminated by traditional chiefs, elders, and other leaders.[2] Additionally, pre-colonial information on the state of the environment through time is contained in documents held by nomadic families (Box 5-1).

During the colonial period, from the mid-eighteenth century until independence in the twentieth century, numerous maps, diaries, travelogs, and descriptive texts were prepared by explorers such as Mungo Park, Stanley, Livingstone, Cameron, Speke, Emin Pasha, Teleki and Du Chaillu (Murray, 1993). After about 1900 the physical and social geography of Africa, as well as its cultural, ethnic, and lin-

[1]For example, comparisons of 1960s maps and 1998 satellite images of Lake Chad show that the lake has decreased in area by approximately 83 percent (USGS, 1998).

[2]Ali Hamma was 110 when he died in 1971. He was a major source of historic information and history for the Zarma people in the region of Loga, Niger.

BOX 5-1
West Africa Legacy Data Examples

- In 1975 the Institut Nigerien de Recherches en Science Humaines published "les Cadres Géographiques à Travers les Langues du Niger—Contribution à la Pedagogie de l'Étude du Millieu." It discusses the views and habits of all the major ethnic groups in Niger and their treatment of factors including space, distance, soil, and rain. The publication demonstrates that each of the main population groups in Niger know and understand landforms, land use and mapping, and that each produces maps of their "terroires" and the common lands surrounding their villages.
- Studies and reports from the colonial period (mid-eighteenth to twentieth century) can be found in such places as the Musée de Tervuren in Belgium, the Musee de l'Homme and the Institut Géographique National (IGN, 2002) in Paris, and the British Museum in London (British Museum, 2002). For example, the IGN has pre-World War I topographic maps of Senegal coastal areas that also depict vegetation and faunal assemblages and historical land-use and land-cover information.
- An archive of historical geographic data is held in trust by a number of nomadic families and clans of Mauritania, including notebooks, parchments, and notes dating to before the tenth century (Ahmed Saleck ben Mohamed Lemine ibn Bouh, 2001). Most of the documents are in the libraries of these families. These documents, which were often updated daily, describe such aspects as weather; availability and location of water; size, type, and location of dunes and other landforms; soils types; towns and villages; and the herding routes taken by the family. Such documents yield valuable information on the resource base, climate change, and the evolution of landscape.
- Most of Africa was mapped by colonial governments beginning in the eighteenth century. During the twentieth century the Belgians, Portuguese, French, and English mapped the continent at various scales. For example, between approximately 1955 and 1965 the IGN mapped French West Africa at a scale of 1:200,000 using aerial photographs. In some cases these maps have been updated as recently as 2001 (e.g., Institut Géographique du Niger/Projet Gestion Resources Naturelles). Britain mapped her colonies at a scale of 1:250,000 using similar techniques. Additionally, there are detailed maps at 1:50,000 (or even 1:10,000) for nearly all of the major cities and most of the river valleys in West Africa (e.g., the Senegal, the Gambia, and parts of the Niger rivers).

guistic context, was described by European[3] and African[4] authors in a more formal, scientific manner to support European competition for economic hegemony. Often forgotten or lost in African countries, these monographs and reports reside in Europe and elsewhere.[5]

During the post-colonial period from about 1960 remote-sensing techniques including aerial photography were used to produce maps at 1:200,000-scale for the Francophone and Lusophone countries and 1:250,000 for Anglophone countries.[6] Unfortunately, legacy data were lost during this period, and work conducted in Africa did build incompletely on the legacy of previous studies and the knowledge of African societies.

Traditionally, African populations, who know a great deal about their environments, have improved their lifestyles, managed their land, and coped with adversity such as drought. Bridges between local knowledge and modern technology are built through the use of legacy data. For example, modern geographic information technologies are of little use in development unless local participants use them to support local activities. If farmers, herders, scientists, and other decision-makers see how the latest technology can benefit their culture, their willingness to learn about and accept such technology can likely be increased through integration of legacy data.

[3]Theodore Monod, who worked in most of West Africa; G. H. Gouldsbury, who described the Great Plateau of northern Rhodesia; J. T. Last, who described the iron workings of the Wa-Itumba in east-central Africa in 1883; C. P. Lucas, who described the historical geography of the British Empire; and Major F. G. Guggisberg who produced 70 maps of the Gold Coast at a scale of 1:125,000 in 1909.

[4]For example, Sekou Toure in Guinea, Leopold Senghor in Senegal, Sir Offori Atta in Ghana, Ibn Fartua in Nigeria.

[5]For example, the Library of Congress in Washington, D.C., has many African holdings. The U.S. Geological Survey Cartographic Library holds the following types of maps: topography, soils, hydrology, geology, ethnicity, minerals, and land use. Additionally, the International African Institute in London and its French counterpart, Institut International Africain, in Paris, have published many African monographs (e.g., the Ethnographic Survey of Africa and the Monographies Ethnologiques Africaines).

[6]The U.S. Central Intelligence Agency prepared a series of African maps at a scale of 1:250,000 that were declassified in the mid-1960s. They were removed from circulation in the mid-1980s when terrorism became a problem. In the 1960s and 1970s image maps and orthophotos of former Portuguese colonies were produced by the Portuguese Army Map Service. Angola and Mozambique were mapped at 1:200,000, Cape Verde Archipelago at 1:25,000, and the Republic of Guinea Bissau at 1:50,000 (ASPRS, 2002).

Unfortunately, knowledge is being discarded as new technologies and methods are introduced in Africa. This loss compromises efforts to address Agenda 21 issues. For example, unaware of the existence of indigenous (Hawsa) maps (Donaint, 1975), donors have not integrated them into their new maps. Consequently, villagers, who are convinced that donors must know best, consider the traditional maps irrelevant. Similarly, children who attend European-style schools in Niger are not taught traditional information on climate and weather. In Mauritania the younger generation of the Maraboutic clans is unaware of the existence of documents describing hundreds of years of their family histories hidden in trunks and libraries. Scientists and NGOs also neglect legacy data. The committee found that the majority of references in recent reports from the World Bank, USAID, and CGIAR group on environment, sustainability, development, and even agriculture, botany, and ethnology postdated 1990.

Efforts are underway to preserve legacy data and ensure that they are used. For example, the Institut Français de la Recherche Scientifique et Technique pour le Development en Cooperation maintains a high profile in protecting and using legacy data through its offices and scientists in French West Africa. Its maps, reports, aerial photographs, and statistical data provide coverage for over 35 percent of Africa. Regional institutes such as Institut Fondamental de l'Afrique Noire in Dakar, Senegal, and the University of Ibadan in Nigeria also have a wealth of environmental and sociological information. Lastly, the DEVECOL database (DEVECOL, 2002) contains results from development initiatives from the past 15 to 20 years in sub-Saharan Africa. In addition to information about soils and climate, the compact-disk-based database documents development experiences in agriculture, soil and water conservation, forestry, biodiversity conservation, drought response, and other rural initiatives.

To complement these efforts to preserve and enhance the use of valuable legacy data, **U.S. government agencies (e.g., USAID and USGS) should assist African countries and organizations to identify, integrate, and maintain existing sources of information (legacy data). They should also provide African countries with copies of such legacy data as reports, maps, statistics, aerial and satellite photographs, and other relevant data and materials currently held outside those countries.** The first task would be substantial, whereas the second would be more routine once the first is addressed.

FRAMEWORK FOUNDATION GEOGRAPHIC DATA FROM MODERN SOURCES

Four framework foundation layers of geographic data are fundamental to at least 20 Agenda 21 issues (Table 2-5): geodetic control, digital elevation, digital ortho-imagery, and human population distribution.

Framework Foundation Layer No.1: Geodetic Control

Geodetic control provides the common coordinate system to which all geographic data are linked. It is used to locate objects and features in terms of their geographic position.[7] Geodetic control can be measured with traditional in situ surveying instruments but this is time-consuming, expensive, and often difficult. One solution is to use a constellation of global positioning satellites. Few nations have the financial and technical resources to launch and maintain such a constellation, much less disseminate the geographic positioning information freely. However, the U.S. Navigation Satellite Timing And Ranging (NAVSTAR)[8] Global Positioning System (GPS) fulfills these goals. Relatively inexpensive,[9] GPS can be used worldwide in a range of applications (Box 5-2). Accurate positioning information, previously impossible for some countries to obtain and difficult to obtain from paper maps that are commonly at 1:50,000 scale or smaller (e.g., Taylor, 1997) can now be collected with a simple hand-held device. Thus, practitioners, scientists, and the general public now have unprecedented access to location information.

The Global Positioning System

The NAVSTAR GPS (Table 5-1; Annex Box 5-1) consists of 24 GPS satellites operated by the U.S. Department of Defense (Parkinson, 1994) (Figure 5-1). At least four satellites are visible at any unobstructed location[10] on the surface of Earth at any time of day. A person with a hand-held GPS receiver may obtain geographic information for any location to within a few centimeters or meters if several conditions are met (Annex Box 5-1).

Because NAVSTAR was initially developed for military use, it is possible for the U.S. government to selectively restrict access to the system, or to adjust the precision of the measurements by introducing a bias, often referred to as "selective availability." Fortunately, selective availability was turned off on May 1, 2000, allowing users worldwide to obtain GPS positioning information without intentional degradation of the information. This situation could change in response to unforeseen events (e.g., war or terrorist activities). The horizontal accuracy of absolute GPS positioning (see Annex Box 5-1) was about 100 m when selective availability was on. Without selective availability the accuracy improved to 5-15 m.

[7]Geographic position includes the horizontal coordinates, as well as elevation.

[8]NAVSTAR is an orbital satellite radio-positioning and time-transfer system (USGS, 1999).

[9]Modest, yet powerful GPS for way-finding cost under U.S.$200. Sophisticated GPS instruments used for professional surveying cost several thousand dollars.

[10]Obstructions might include mountains, tall buildings, or dense multiple-story vegetation.

TABLE 5-1 Characteristics of the NAVSTAR Global Positioning System of Particular Significance for Developing Countries

- Geographic positioning accuracy from meters down to centimeters.
- Geographic information is provided in 3 dimensions (latitude, longitude, and elevation).
- Signals are available to people anywhere on Earth (air, land, or sea) without discrimination.
- Signals are available free of charge to anyone. However, the user must possess a GPS receiver. GPS receiver costs continue to decline.
- Available 24 hours a day, 7 days a week.
- It is an all-weather system not affected by clouds (however, intense rain or thick vegetation canopy can reduce its effectiveness).
- Geographic coordinates are tied to a single global geodetic datum.
- No inter-station visibility is required for precise positioning. This means that it is not necessary for a surveyor to use a theodolite to view a distant stadia rod.
- Geographic position can be determined rapidly (in seconds to minutes).

SOURCE: Adapted from Rizos (2002).

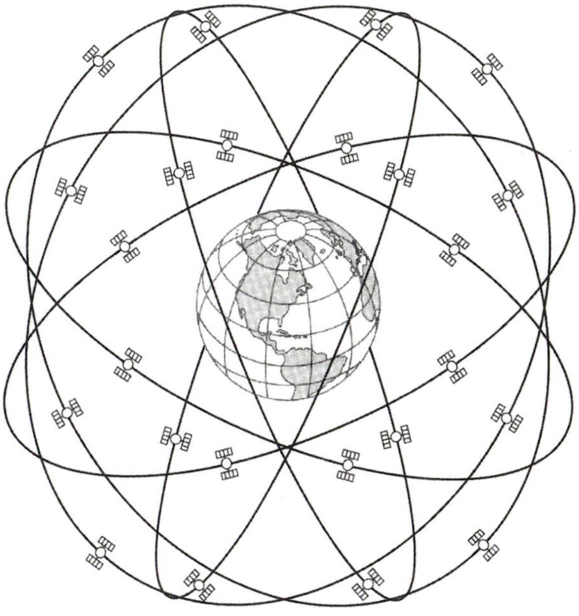

FIGURE 5-1 The NAVSTAR GPS consists of a constellation of 24 GPS satellites located at approximately 20,200 km above Earth's surface. SOURCE: Courtesy of Rizos (2002).

GPS information is broadcast worldwide to virtually anyone in any country and is of great importance to the practical collection and use of fundamental geographic data for Agenda 21-related initiatives. Its utility should not be reduced by reintroducing selective availability, and its continuity should be guaranteed. The U.S. Department of Defense should continue to allow free access to Global Positioning System data.

In March 2002 European leaders announced that they were pushing ahead with the ambitious and sometimes controversial Galileo satellite positioning system. As proposed, the system will comprise 30 satellites in orbit by 2007 and will provide a position and time service that both complements and competes with the U.S.'s GPS constellation. Galileo will be the third global navigation satellite system, joining Russia's GLONASS (Global Orbiting Navigation Satellite System) and the U.S. Global Positioning System (Barnes, 2002). Users in Europe, North America, and around the world will benefit if Galileo is designed and built so that it is interoperable with GPS.

Framework Foundation Layer No.2: Ortho-imagery

An ortho-image is prepared from an aerial photograph or remotely sensed image. It has the metric qualities of a traditional line map with the rich detail of an aerial image.[11] Com-

[11]During processing an ortho-image is adjusted (rectified) to a standard map projection and datum. Geometric errors caused by topography and other anomalies are removed from the data set during processing (Thrower and Jensen, 1976; Jensen, 1995).

BOX 5-2
Applications of GPS

GPS has a number of applications, including

- locating villages and dwellings in censuses (e.g., in the 2000 Niger census);
- mapping boundaries such as national borders (e.g., to resolve border disputes) or the extent of private land holdings (e.g., to build a cadastral database); and
- tracking the movements of wildlife (e.g., elephants).

The Save the Elephants project operates in many African countries and relies on GPS to track the movements of elephants (ESRI, 2002). The project goals are to assist wildlife departments in combating poaching, educating people globally about elephants, and learning more about elephant behavior. An elephant's position is tracked using a collar with a GPS receiver. Its location is recorded every one to three hours and collected every few months by flying over the elephant and downloading the data onto a laptop computer. The travel patterns of the elephant are subsequently mapped using a GIS. The fine detail afforded by the GPS data allows researchers to learn about decisions made by elephants on their needs for food, water, and safety. Safe corridors for elephant travel are determined with this information, and human and ecological factors are weighed by wildlife managers to reduce conflicts between humans and elephants.

monly in developed countries, ortho-images are obtained from high spatial resolution (Annex Box 5-2) aerial photography.[12] Ortho-images can also be produced from high spatial resolution satellite imagery such as DigitalGlobe panchromatic[13] data (61 × 61 cm). High spatial resolution aerial photography or satellite imagery is expensive, however, and therefore often impractical for users in developing countries. The Landsat Thematic Mapper Global GeoCover-Ortho database is a valuable alternative for users in developing countries.

Landsat Global Database

The Global GeoCover-Ortho database is a color ortho-image database with a spatial resolution of 30 × 30 m covering the majority of Earth's landmass. In addition to being a data source for assessing global land cover and modeling global climate change, it can be a base map for natural resource managers and development planners in the developing world.

The database originated from a 1998 contract between NASA and Earth Satellite Corporation (EarthSat) as part of the NASA Scientific Data Buy program (Box 5-3). The majority of the data was acquired by the Landsat Thematic Mapper (TM) remote-sensing system (Annex Box 5-3). Consequently, the GeoCover-Ortho images are the most accurate commercially available satellite-derived base maps of the world. With a positional accuracy of less than 50 m (root mean square error), they are more accurate than most of the world's 1:200,000-scale maps. Furthermore, owing to the nature of the original contract set up by NASA, this imagery is more economically accessible for developing countries. It is a comprehensive global data set with image dates ranging from 1987 to 1993 (Annex Box 5-4), and is uniquely suited to establishing a worldwide environmental baseline. Additional Landsat TM images (and many other types of remotely sensed data) can be overlaid on the GeoCover-Ortho imagery for purposes of change detection.

Three types of GeoCover-Ortho products are available from EarthSat (<http://www.earthsat.com>) individual images (e.g., Figure 5-2), mosaics, and regional mosaics (e.g., Figure 5-3) (Annex Table 5-1).

Earth Satellite, Inc., is using imagery collected during the GeoCover-Ortho project to map and classify global land-cover change over a decade using the 1990 dataset in comparison with a 2000 data set (Chapter 6).

Framework Foundation Layer No.3: Digital Elevation and Bathymetry

Elevation data (often referred to generically as topographic data) provide information about terrain. These data are used in many different applications (Table 5-2). Unfortunately, owing to the inaccessibility of many of Earth's mountain chains, deserts, and forests, elevation information about such areas can be inaccurate, of limited extent, or nonexistent. Even where access is practical, traditional surveying methods are expensive. Furthermore, neighboring countries may use differing data-collection methods that cause data discontinuities at borders, whereas natural resources (e.g., rivers) often cross these borders. To address these challenges the United States has partnered with a number of countries and organizations to produce two digital elevation datasets: the GTOPO30 (Global Topography at 30 arc seconds)

[12]This aerial photography may be black and white panchromatic, color, or color-infrared. For example, the Digital Orthophoto Quarter Quad program of the U. S. Geological Survey is based on 1:40,000-scale National Aerial Photography Program color-infrared aerial photography processed to 1 × 1 m spatial resolution.

[13]Panchromatic images are created by recording reflected energy over a relatively broad portion of the electromagnetic spectrum (e.g., 0.5 – 9.0 μm).

BOX 5-3
NASA's Scientific Data Buy Program

In 1997 the U.S. Congress allocated $50 million to NASA for the Scientific Data Buy, a demonstration program encouraging NASA to purchase remotely-sensed data from the private sector. NASA's John C. Stennis Space Center in Mississippi administered the program.

NASA established partnerships with six companies (AstroVision; DigitalGlobe; EarthSat; EarthWatch, Inc. [now DigitalGlobe]; Positive Systems, Inc.; and Space Imaging) to participate in the Scientific Data Buy. Four research themes addressed with data from these companies include land-cover and land-use change; seasonal-to-interannual climate variability; natural hazards; and long-term climate change (NASA, 2002). Data are made available to NASA-funded scientists through a competitive selection process.

The Scientific Data Buy program exemplifies the changes under way in the existing relationship among U.S. federal science agencies, the private sector in the form of the commercial remote-sensing industry, and both scientific and applied users of remotely-sensed data (NRC, 2001). A forthcoming NRC report (*Toward New Remote Sensing Partnerships: Government, the Private Sector, and Earth Science Research* [in editing]) examines the issues of institutional relationships between the government, industry, and the research community for remote-sensing.

SOURCE: NRC (2001), NASA (2002).

FIGURE 5-2 A GeoCover-Ortho image of Mount Kilimanjaro originally obtained at 30 × 30 m spatial resolution. It has a positional accuracy of better than 50 m (root mean square error). Landsat TM bands 7,4,2 (mid-infrared, near-infrared, and green) are displayed (courtesy of Earth Satellite Corporation). The width of the depicted area is 110 km. Each color or shade is unique and depends on the vegetation type, health, and growth stage. The bright greens are dense vegetation. The hot reds in the bottom left section are crops. The purples and pinks are sparse to no vegetation. The bottom third center of the image along Mount Kilimanjaro's lower slopes contains areas of clear cuts (in pinks) surrounded by uncut verdant forest (bright greens). The top of the mountain is snow-covered and the white areas are clouds.

dataset, and the 2000 Shuttle Radar Topography Mission (SRTM) dataset. The GTOPO30 dataset is a global digital elevation dataset, and the SRTM dataset covers 80 percent of the global land mass.

Global Topography at 1 × 1 km Spatial Resolution

The GTOPO30 global digital elevation data set (Figure 5-4) has a spatial resolution of approximately 1 × 1 km (Gesch et al., 1999; GTOPO30, 2002). Completed in 1996, it was developed through a collaboration led by the U.S. Geological Survey's EROS Data Center.[14] Diverse sources of topographic information were used to produce the GTOPO30 data set (Figure 5-5). GTOPO30 data for Africa were derived manually by converting topographic information from the Digital Chart of the World (USGS, 2002a) into digital format and supplementing this with digital elevation data.

A version of the GTOPO30 data set for Africa was released as part of the *Global GIS Database: Digital Atlas of Africa* in 2001 by the U.S. Geological Survey. It is available on a single compact disk at a cost of about U.S.$10 (USGS, 2002b). The database is viewed using ESRI's ArcView Data Publisher software, also on the same compact disk (Hearn et al., 2001).

Although the GTOPO30 dataset offers global coverage, it is of limited value in Africa and most other developing countries for monitoring ecosystems, urban and rural infrastructure, and hydrology because of its coarse resolution.

[14] The following organizations participated by contributing funding or source data: NASA, the U.N. Environment Programme/Global Resource Information Database (UNEP/GRID), USAID, the Instituto Nacional de Estadistica Geografica y Informatica of Mexico, the Geographical Survey Institute of Japan, Manaaki Whenua Landcare Research of New Zealand, and the Scientific Committee on Antarctic Research.

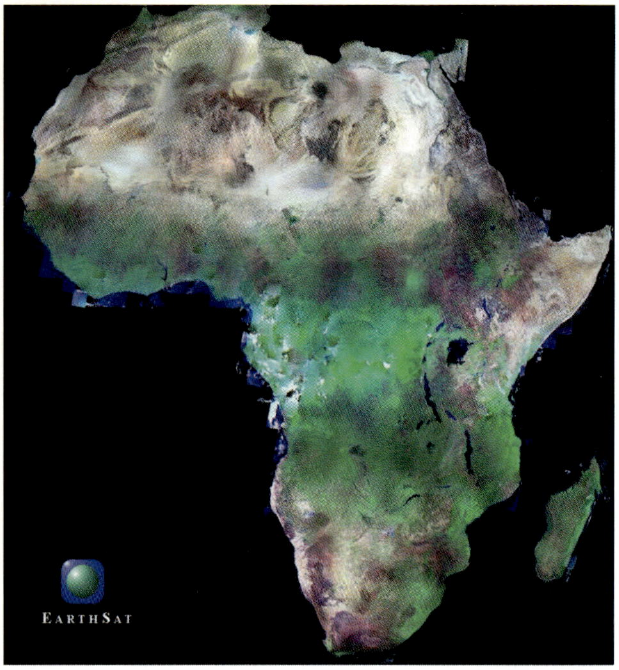

FIGURE 5-3 A GeoCover-Ortho regional mosaic of hundreds of Landsat Thematic Mapper images of Africa obtained between 1987 and 1993 at the height of the growing seasons and with very little cloud cover. TM bands 7,4,2 are displayed. Individual images used to produce this mosaic represent the most accurate satellite imagery available for the entire continent of Africa (courtesy of Earth Satellite Corporation.). The distance (parallel to the Equator) from the tip of Horn of Africa to the west coast off Guinea-Bissau is 7,350 km.

TABLE 5-2 Examples of Users and Applications of Elevation Data

Occupation	Uses of Elevation Data
Natural resource managers	Monitoring and management of soil erosion and biodiversity; deduction of soil type, or likely occurrence of floral and faunal habitats (from slope and aspect information).
Urban/rural planners	Determining suitable locations for structures, transportation networks, wetland protection, and other land-use planning.
Pilots	Flight planning and navigation.
Engineers / hydrologists	Designing safe and efficient hydrologic projects such as dams and levees and to mitigate problems due to flooding.
Geologists	Mitigating landslide and earthquake hazards; volcano monitoring.
Communications	Optimizing location of transmitters through knowledge of height and location of natural and human-made obstacles.
Military	Planning and simulations during training and real-time operations.
Tourism	Navigation.

Global Topography at 90 × 90 m Spatial Resolution (or Better)

The SRTM data set is superior to GTOPO30 because of its higher spatial resolution[15] and in its uniform origin.[16] Because the dataset is from a homogeneous source referenced to a uniform global geodetic datum, users will be working from a common reference frame.

The SRTM data-gathering mission (Box 5-4) was an international project involving the U.S. National Imagery and Mapping Agency (NIMA), NASA, the German Aerospace Center (Deutsches Zentrum für Luft und Raumfart), and the Italian Space Agency (Agenzia Spaziale Italiana), and is managed for NASA by the Jet Propulsion Laboratory, California Institute of Technology. The objective of the mission was to obtain elevation data at 30 × 30 m spatial resolution with a near global coverage and generate the most complete high-resolution digital topographic database of Earth.[17] The mission collected data from over 80 percent of Earth's landmass, home to nearly 95 percent of the world's population (Figure 5-7). The increased topographic detail of SRTM data over GTOPO30 data is apparent in Figures 5-8 and 5-9, which show data for the Mount Kilimanjaro area in Tanzania and Kenya.

The raw SRTM data are being processed into a digital elevation model by NASA's Jet Propulsion Laboratory (An-

[15]This dataset is at a minimum a factor of ten more detailed than GTOPO30 and matches the resolution of Landsat GeoCover-Ortho database discussed above.

[16]It was produced using a standardized, accurate technology during a single 11-day collection period beginning February 11, 2000.

[17]Other details of the mission objectives included better than 16-m absolute vertical accuracy, better than 10-m relative vertical accuracy, and better than 20-m absolute horizontal accuracy. All accuracies are quoted at the 90 percent level, consistent with U.S. National Map Accuracy Standards.

BOX 5-4
The Shuttle Radar Topography Mission

During its 11-day mission the SRTM acquired enough data to produce the most accurate, homogeneous, detailed, and complete digital elevation model of Earth's land surface ever constructed. Surface elevation is calculated using radar interferometry, wherein differences are compared between two radar images taken from slightly different locations (see Figure 5-6) (SRTM, 2002b).

Radar is an excellent sensor for measuring topography because it can operate both day and night and it is unaffected by clouds. Radar interferometry yields accurate topographic data unless the vegetation canopy is extremely dense, in which case the elevation values may be biased toward the canopy height rather than the actual terrain elevation.

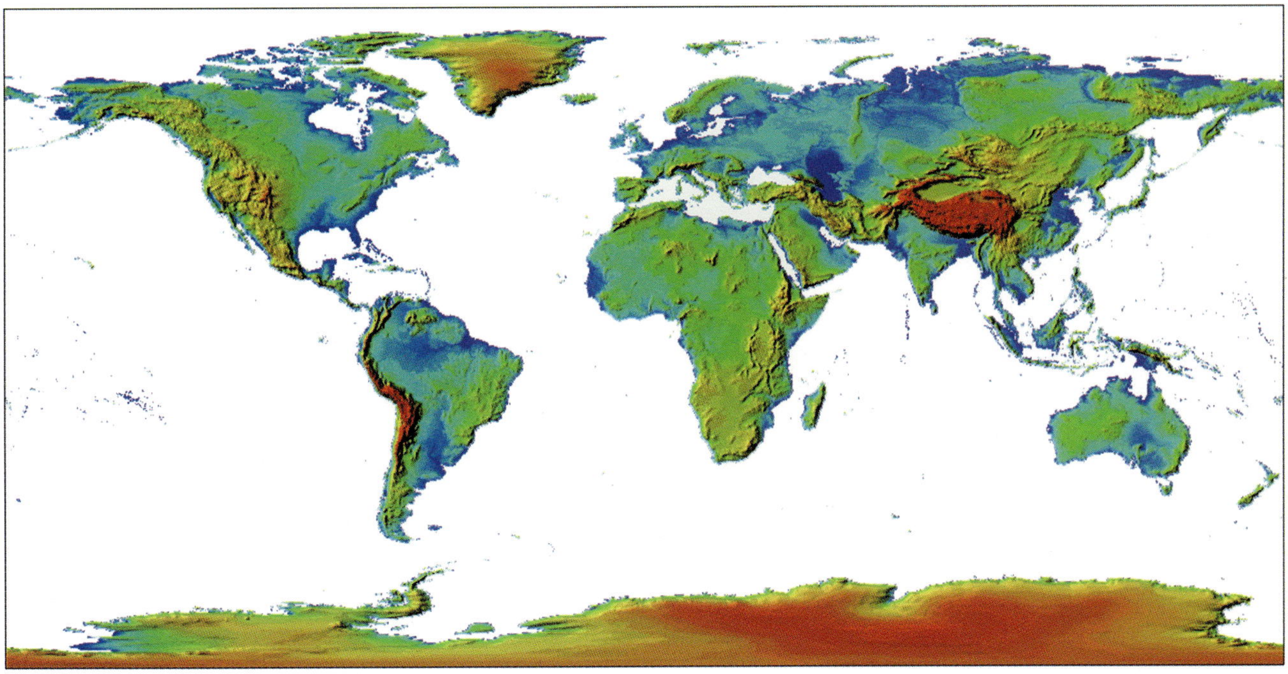

FIGURE 5-4 Global elevation data contained in the GTOPO30 digital elevation dataset. The spatial resolution is 1 × 1 km (U.S. Geological Survey EROS Data Center). The dataset covers the entire globe.

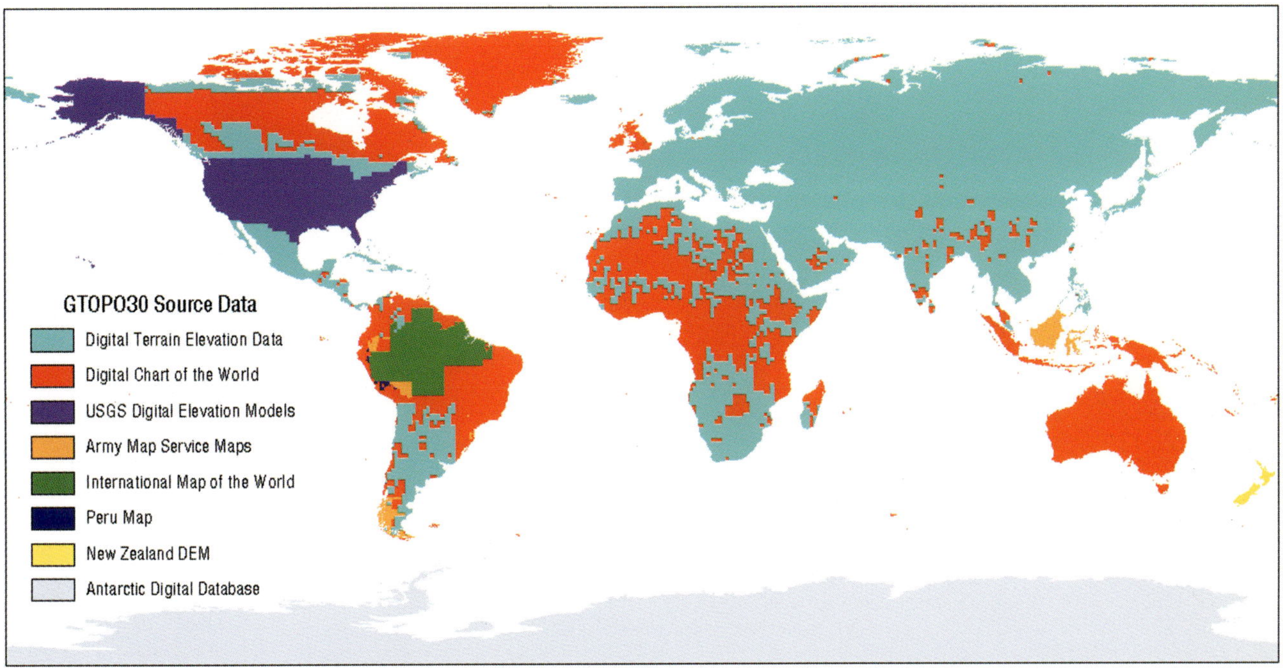

FIGURE 5-5 Sources of topographic information used in the production of the global GTOPO30 digital elevation dataset. A tremendous variety of sources, both traditional maps and digital elevation data, were used. Much of the data for Africa was derived by digitizing information from the Digital Chart of the World (1:1,000,000-scale), which does not contain very detailed topographic information (U.S. Geological Survey EROS Data Center).

FIGURE 5-6 Artist's rendition of the SRTM 60-m (200-ft) mast being deployed from the space shuttle *Endeavor* (courtesy NASA Jet Propulsion Laboratory). Radar images are collected from the end of the mast and from the shuttle payload bay.

nex Box 5-5) (SRTM, 2002a), and the products can be tailored to meet the needs of civil, scientific, and military users. In the current plan, which is not finalized, SRTM data will be released at 30 × 30 m spatial resolution for the United States and at 90 × 90 m spatial resolution for the rest of the world. **NASA should produce digital elevation data from the SRTM at the highest possible spatial resolution (e.g., 30 × 30 m) for all areas. The data should be made available without restriction and at affordable cost. NASA should also provide the synthetic aperture radar orthoimage mosaics at 30 × 30 m spatial resolution that are being produced as part of the processing.** These mosaics would provide additional information about land-cover conditions and surface roughness characteristics especially in tropical regions perennially shrouded by cloud cover.

Framework Foundation Layer No.4: Human Population Distribution Information

National census data provide the foundation for measuring population distribution and change at the national to lo-

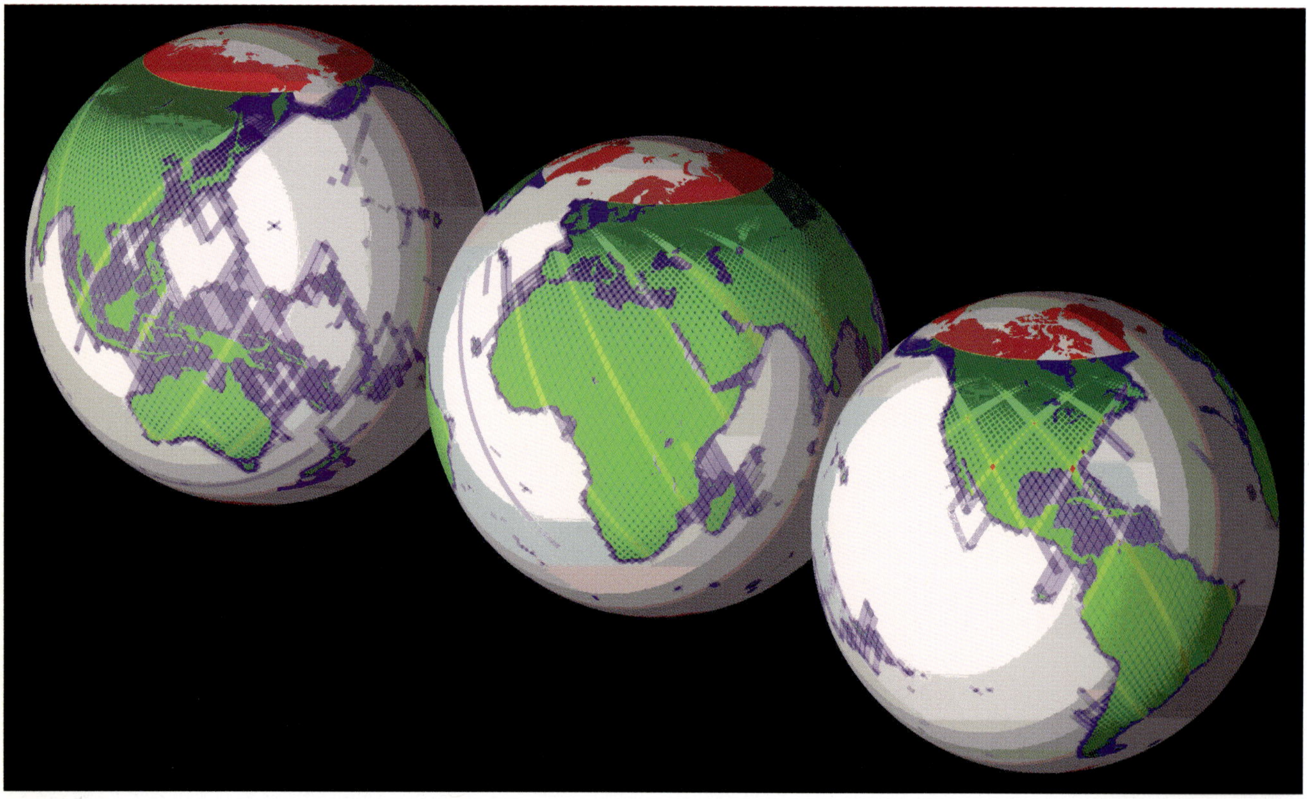

FIGURE 5-7 Global coverage map of Shuttle Radar Topography Mission data collection. Most of the covered area is between latitudes of 60 degrees north and 56 degrees south. Data were mostly acquired over land (indicated in shades of green), with small amounts of data collected over the water for calibration purposes (shades of blue). Of the mapped area, 99.97 percent was covered once (green), 94.59 percent was covered twice (yellow-green), 49.25 percent was covered three times, and 24.10 percent was covered four times. Areas in red could not be mapped because of time constraints (50,000 km^2 was not covered, all within the United States) (courtesy NASA Jet Propulsion Laboratory).

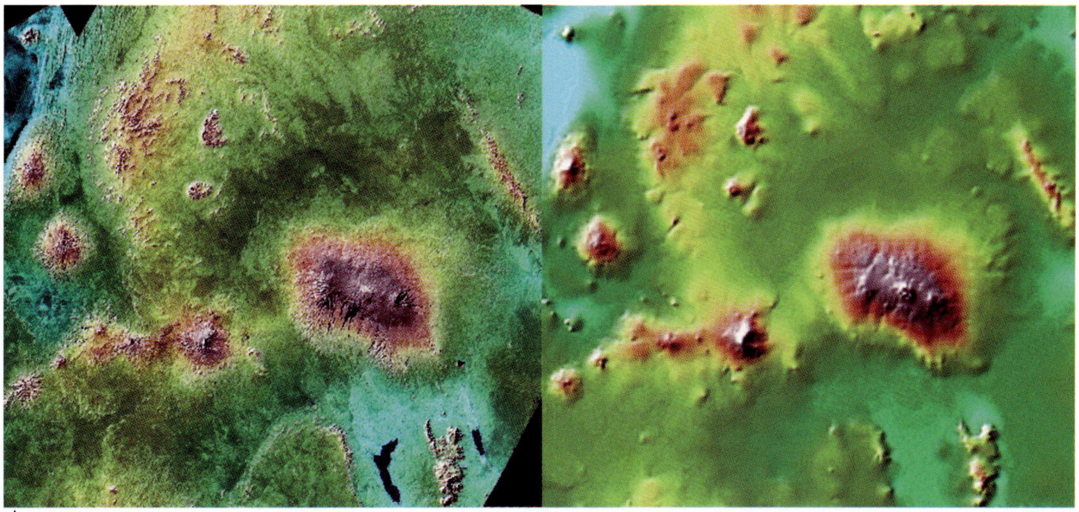

A B

FIGURE 5-8 (a) SRTM 30 × 30 m digital elevation model of the area centered on Mount Kilimanjaro in Tanzania and Kenya. (b) GTOPO30 1 × 1 km data of the same region. GTOPO30 data fail to identify major lakes in the region (NASA Jet Propulsion Laboratory and U.S. Geological Survey EROS Data Center). The distance between the two peaks is 70 km.

A B

FIGURE 5-9 (a) SRTM 30 × 30 m digital elevation model of Mount Kilimanjaro in Tanzania and Kenya observed from an oblique vantage point looking W-NW. Note the detail associated with the small cinder cones on the flank of the mountain. (b) GTOPO30 1 × 1 km data of the same region (NASA Jet Propulsion Laboratory and U.S. Geological Survey EROS Data Center). The distance between the two peaks is 70 km.

cal levels (Liverman et al., 1998). The strengths of human population censuses arise from their completeness of coverage; continuity of statistics from census to census; and the detail that each census provides about population sub-groups in local areas. In the current worldwide development arena such key issues as good governance, anti-poverty strategies, and the need to promote economic growth with social equity all require population and other demographic[18] data at the detailed local scale that only a population census can provide. Moreover, there exists an increasing demand for disaggregated data at the sub-national level. Data gaps are inevitable without a recent census. Without a recent census, surveys must use outdated sampling frames, administrative boundaries could be incorrectly drawn, and national and sectoral planning and decision-making will be based on outdated and unreliable statistics. Even basic data on population size and age composition will be unavailable or unreliable.

This section reviews the current status of human popula-

[18]For example, data on fertility, mortality, migration, age structure, and household composition.

tion censuses in Africa and discusses how these can be integrated with other geographic data types. It also reviews tools for rapid access to estimates of population distribution during humanitarian crises: the LandScan 2000 global population estimate and the Gridded Population of the World. The section ends by reviewing how remotely sensed data can be used to estimate population distribution.

African Demographic Censuses

Human population data for Africa are unreliable because of the dearth of good demographic censuses. Seventeen African countries did not conduct their first modern population census until the 1980 census round (i.e., between 1975 and 1984). In recent years civil unrest has taken its toll on census-taking in a number of African countries, as has economic stagnation and higher priority issues (e.g., HIV/AIDS).

Ideally, countries should conduct population censuses at 10-year intervals (UN, 1998).[19] The 2000 round of censuses (1995 to 2004) has encountered a number of challenges, some new and others longstanding (including funding, politics, civil unrest, nomadic populations). Funding constraints have affected the 2000 census round. Censuses today cost much more than in the past, partly because of increasing population size[20] (Figure 5-10), and partly because of a growing demand for more specialized and detailed information. In addition, despite the rising costs of censuses, there appears to be "donor fatigue" in meeting gaps in census funding. For example, the United Nations Population Fund's current policy is to limit support to countries conducting their first or second population census (Leete, 2001).

Recently, a number of bilateral and multilateral agencies (e.g., USAID, U.K. Department for International Development [DfID], CIDA, the World Bank, the European Union, UNDP, UNICEF, and the African Development Bank), as well as governments (e.g., Belgium, Finland, France, Japan, Norway, Portugal, Sweden, and Taiwan), have provided limited assistance to conduct censuses. Donor statistical agencies such as Statistics Canada and the U.S. Bureau of the Census often do not have direct funding from their governments to support international assistance and are reliant on funding from other government agencies, usually the international development agency.

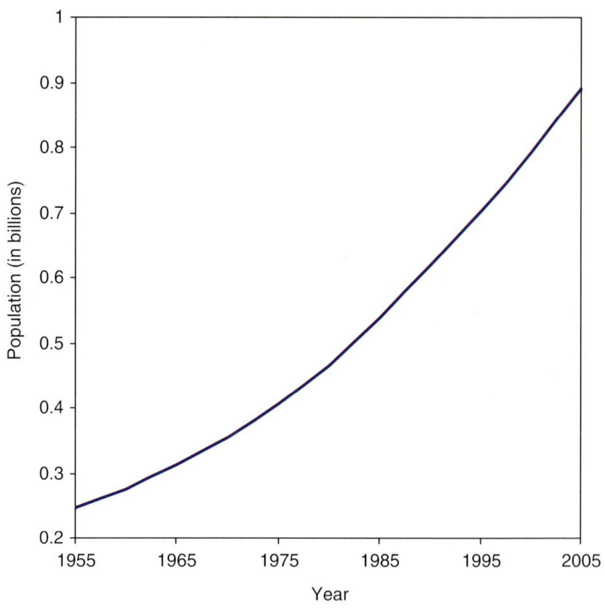

FIGURE 5-10 Population growth in Africa, 1955-2005 (UNDP, 2001).

As of mid-2002 roughly two-thirds of African countries have yet to conduct a population census in the 2000 census round (Annex Table 5 2). Currently, out of 44 countries in sub-Saharan Africa, 35 countries have moderate to severe funding gaps for their 2000 round of census operations (UNFPA, unpublished). Limited funding also restricts the dissemination of censuses. In Burkina Faso, for example, the government had funding to publish no more than about 50 copies of the 1996 census. Unless satisfactory solutions are found, some sub-Saharan countries may find it difficult to organize further population censuses (Diop, 2001).

Even when sufficient funds exist to conduct censuses in Africa, there are additional challenges to their completion, including

- their politicized nature, with the collection and dissemination of census results delayed (or even suppressed);
- difficulties enumerating nomadic and dispersed populations;[21]
- village names changing from one census to another;
- maps on which the villages were noted are commonly out of print (the problem of loss of legacy data);
- census "maps" may be sketches with indeterminate boundaries;

[19]UN (1998) also states that "censuses must be carried out as nearly as possible in respect of the same well defined point of time and at regular intervals so that comparable information is made available in a fixed sequence."

[20]Since the early 1960s, when many African countries began to achieve independence, the population of the African continent has nearly tripled, increasing from some 277 million in 1960 to nearly 800 million in the year 2000. Despite the impact of the AIDS pandemic, Africa is expected to experience significant population growth over the next half century. According to (medium variant) projections prepared by the U.N. Population Division, the population of the African continent will increase to nearly 1 billion inhabitants by 2010 and to over 2 billion by 2050.

[21]For example, Eritrea was scheduled to have a census in March 1997 but, for a variety of reasons, it was postponed until 2002. Because of the delay, over 50 percent of the Enumeration Areas had to be re-demarcated due to population growth, the re-settlement of displaced persons, and the formation of new villages. In doing this, villages, roads and tracks were plotted with GPS receivers (UNFPA, unpublished).

- many years may pass before data at the sub-national level are processed and disseminated to local users, even though data at the local level are important for place-based development planning; and
- although data at the village level are highly useful, data often are aggregated to the district or even to the regional level, masking the critical local interrelationships between population and the environment, health, land use, infrastructure, education, and poverty.

The Value of Geographically Referenced Population Data

Governments need national population data and the international community needs global population data to understand the impacts of population on the environment (and, conversely, to better understand the impact of the natural environment on the well-being, vulnerability, and livelihoods of populations). The challenge lies in linking population and other socio-economic data with information on the environment in a common, geographically referenced framework.

Over the past decade the U.S. Bureau of the Census has conducted advanced work on the collection and presentation of geographically referenced data. Their TIGER (Topologically Integrated Geographic Encoding and Referencing) files contain a digital database of geographic features covering the United States. The TIGER database defines the location and relationship of roads, rivers, railroads, and other features to each other and to the numerous geographic entities for which the Bureau of the Census tabulates data both from its population censuses and its sample surveys. The database provides a convenient way to link to a geographic location the statistical data being reported for these entities.

No developing country has the equivalent of the TIGER database. However, in recent censuses as well as in some surveys African countries have begun to use GPS receivers to record the coordinates of villages (Box 5-5). These GPS-based efforts provide a means of integrating census data with other data sets (e.g., agricultural statistics; location of roads, facilities, villages). GPS is not a "magic bullet," however. In Niger's recent census, for example, it was estimated that GPS data were correct in 85 percent of cases, marginally incorrect in another 5 percent of cases, and completely incorrect in 10 percent of cases, owing to such factors as reversal of latitude and longitude coordinates or insufficient training to operate the GPS receiver.

Estimating Human Populations at Risk

Natural and human-induced disasters place human populations at risk, often with little or no advance warning. Estimates of evolving population distribution are required at short notice for emergency response by national and international organizations. One approach used to estimate populations at risk is to analyze existing population information (as

BOX 5-5
The Value of Geographically Referenced Data: Lessons Learned in Niger

In 1994 USAID financed a pilot study called "Population, Health and Environment in Niger." This study was executed by the government of Niger, CERPOD, and AGRHYMET (Chapter 3). The goal was to develop a GIS-based presentation showing relationships between population, health care, and the environment, using data from the 1988 Niger Census, the National Health Information System, and the environmental sources.

The pilot was to develop and demonstrate a decision-support system for locating health centers in Niger. The system would support the government's planning as it attempted to increase the national health coverage rate from 32 percent to 45 percent of the population. The first step was to integrate existing population and health center data sets into a GIS database. When originally collected, these data had no accompanying geographic coordinates, and could not, therefore, be readily plotted on a map. Project members found that the best source of geographic information for the location of population at the village level was old paper maps (legacy data) produced in the 1950s and 1960s by IGN in Paris (Box 5-1). Information from these maps was entered into the computer database, yielding coordinates of each village on which the population statistics could be plotted. The population data for each village were then manually linked to the IGN base map to show population distribution across Niger. Owing to recurring problems with multiple names for villages, the age differential between the base map and the census data, and incomplete coverage by the legacy maps from IGN, only 87 percent of the population of Niger was covered in the GIS database.

Using the population data along with the location of existing health facilities, the project demonstrated a GIS that could guide government decision-makers in the optimal location of new health facilities (facilities that were accessible to the greatest number of people with the minimum number of new centers). The GIS was never adopted, however. Upon completion of the pilot project the team disbanded and the data and the software were separated. There currently exists no copy of the GIS at CERPOD, for example (H. Wane, CERPOD, personal communication, 2002). Insufficient local capacity was built to support further application, owing to the short duration of the pilot.

There were positive impacts of the pilot project, however. CERPOD convened a workshop in 1996 that demonstrated to Sahelians and others the potential of integrating population data into spatial decision-support tools such as GIS. Organizations, including UNICEF, supported the meeting. Subsequent population censuses by UNICEF, Demographic Health Surveys, and others at the national and local level have routinely collected geographic coordinates in addition to population statistics.

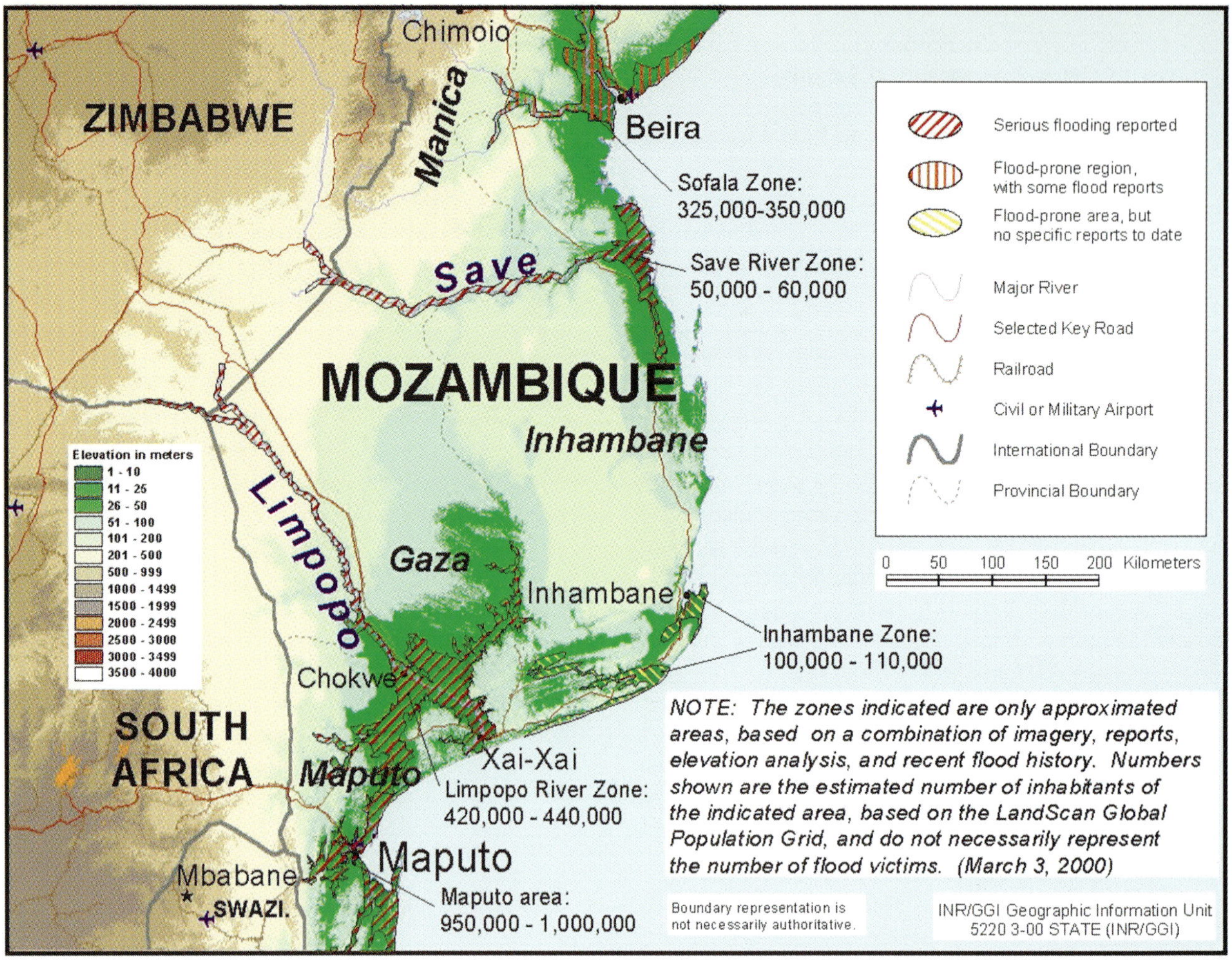

FIGURE 5-11 A flood hazard map showing populations at risk in Mozambique following the cyclone Eline in March 2000 (courtesy of the Geographic Information Unit, U.S. Department of State). It was used to estimate the areas of Mozambique in greatest need of humanitarian assistance following the flooding.

good as can be obtained in a region) in conjunction with cartographic and remote-sensing information.

LandScan 2000

LandScan 2000 is a worldwide population database for estimating ambient populations at risk (e.g., Figure 5-11). Unlike traditional censuses that tie the location of people to their homes, LandScan 2000 attempts to account for people's mobility. It aims to map the presence of people in fields, schools, or on roads, rather than solely in their dwellings. The database was developed as part of the Oak Ridge National Laboratory Global Population Project (Dobson et al., 2000). The types of geographic and other information that enter into the calculation[22] include census data (usually at the provincial level), road networks, elevation distribution, land cover (from AVHRR satellite imagery), and nighttime light sources (as detected from the Defense Meteorological Satellite Program [Box 5-6 and Annex Box 5-6]). LandScan allocates a certain number of people to each 1 × 1 km cell based on the relative likelihood of population occurrence associated with the aforementioned variables. The LandScan files are available free of charge by continent (LandScan, 2000).

Gridded Population of the World

The Gridded Population of the World (GPW) is an estimate of world population for 1990 and 1995. The estimates are based on the nearest population censuses to 1990 and 1995, and are adjusted to agree with UN population estimates for those years for each country. The data are freely available at global, continental, and country levels (CIESIN, 2002). GPW takes the best resolution data available to generate a 5 × 5 km population grid.[23]

[22]A statistical estimate of the likely population in each grid cell is obtained by considering each of the following variables: proximity to roads, slope steepness, land-cover type, and frequency of nighttime lights. The resultant probability field is then linked to the census data to yield an estimate of likely location of the population.

[23]See <ftp://ftp.ciesin.org/pub/gpw/ancillary/GPW2.xls>.

BOX 5-6
Defense Meteorological Satellite Program
Global Nighttime Lights: An Indication of Human Settlement Patterns

Global meteorological and oceanographic conditions, among other parameters, are monitored by the U.S. Air Force's Defense Meteorological Satellite Program (DMSP). One sensor system on DMSP satellites (the Operational Linescan System [OLS]) was developed originally to monitor the global distribution of clouds and cloud-top temperatures. In addition to these applications, however, the OLS detects faint sources of light at night on Earth's surface, including settlements and fires (Welch, 1980; Elvidge et al., 1996, 1997, 2002) (Figure 5-12).

By defining a reference set of stable lights—those present in the same location on a consistent basis—it is possible to identify new settlements or expansion or contraction of existing settlements by comparison with images collected at a later time.

TABLE 5-3 A Comparison of Features of the LandScan 2000 and Gridded Population of the World Global Population Estimates

Features	LandScan 2000	Gridded Population of the World
Number of administrative units	69,350[a]	127,093[a]
Resolution (grid-cell size)	1 × 1 km	5 × 5 km
Minimum downloadable unit	Continent	Country
Model input variables	Census data (population) Land cover Roads Elevation Nighttime lights	Census data (population)
Population data source	U.S. Bureau of Census	Varied data sources

[a]Roughly 61,000 of these administrative units are within the United States.
SOURCE: CIESIN, 2002

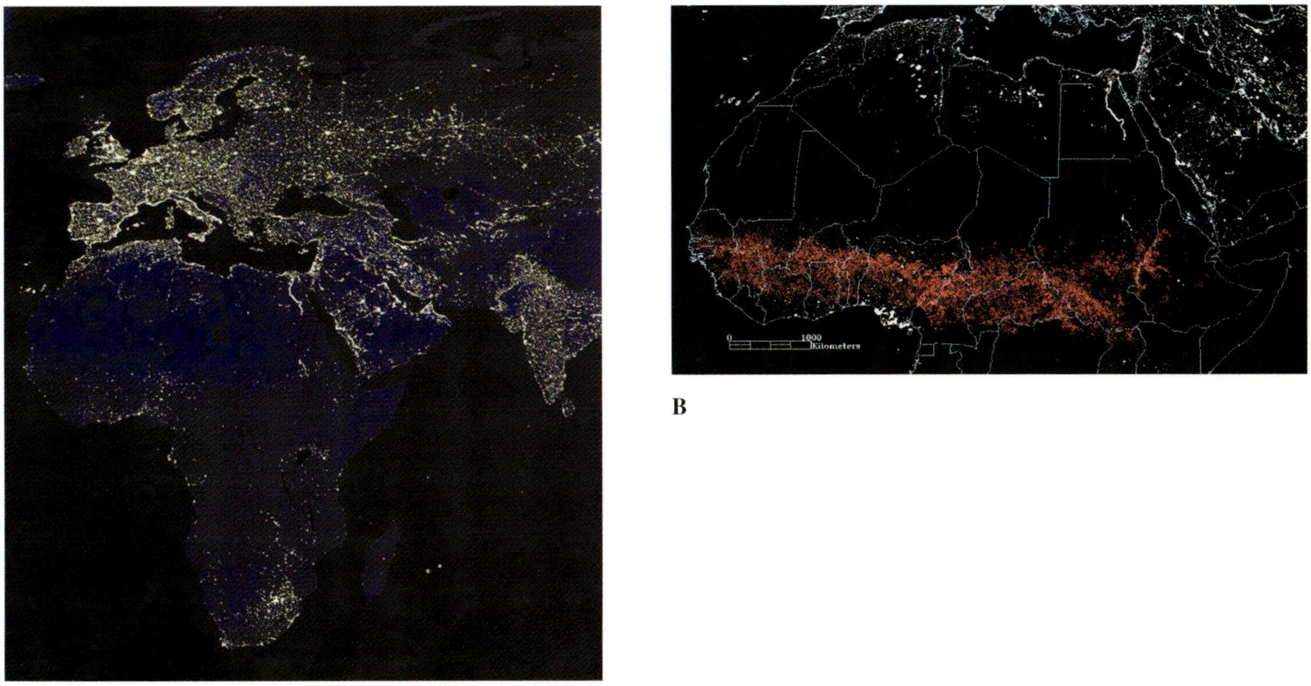

A

B

FIGURE 5-12 (a) The distribution of settlement lights derived from Defense Meteorological Satellite Program Operational Linescan System for Africa (courtesy Defense Meteorological Satellite Program). The distance (parallel to the Equator) from the tip of the Horn of Africa to the west coast off Guinea-Bissau is 7,350 km. (b) Distribution of fires in sub-Saharan Africa (courtesy Defense Meteorological Satellite Program).

By using census data from many more administrative units outside the United States, GPW provides a more detailed estimate of global baseline population distribution than LandScan (Table 5-3). Users can select the most reliable among varied data sources (e.g., from national statistical offices or the United Nations). Unlike LandScan, however, data in GPW have not been adjusted to account for people's mobility.

Obtaining Human Population Distribution Directly From Remotely Sensed Imagery

There is no substitute for a ground-based, geographically referenced population census, but they are expensive and time-consuming to conduct. In the search for alternative approaches there is interest in estimating population distribution using remote sensing-based models (Liverman et al., 1998). If sufficiently accurate in situ data are available to calibrate these models, their accuracy may approach that of traditional censuses. Such population estimation can be performed on (1) counts of individual dwelling units at the local level, (2) measurement of total urbanized land areas (often referred to as settlement size) at the sub-national level, or (3) estimates derived from land-use/land-cover classification at the national level (Lo, 1995; Sutton et al., 1997) (Box 5-7).

Progress toward Agenda 21 goals is impeded by a lack of complete, reliable data on human population distribution. Currently, many African countries are challenged to complete the present round of censuses, in part because of funding constraints. These data will be of greatest value to decision-makers (including those using remote-sensing techniques) if they are geographically referenced and are as disaggregated as possible. **USAID and the U.S. Bureau of the Census should provide financial and technical support to national census offices and bureaus in Africa to help them complete censuses, geographically reference the data, and make the data available in disaggregated form to decision-makers.**

FRAMEWORK THEMATIC GEOGRAPHIC DATA

Upon the framework foundation data layers are placed framework thematic databases. This section reviews existing or potential sources of the four critical framework thematic databases: hydrology, government boundaries, transportation, and cadastre.

Hydrology

Hydrologic framework thematic data underlie most natural resource, urban infrastructure, and utility planning applications. They can be obtained using a variety of geographic data technologies and sources. First, they can be recorded in situ by people with GPS. Second, they can be extracted from maps or remotely sensed data (e.g., in Burkina Faso–see Chapter 8). Third, stream network information can be extracted from digital elevation models.

One product is HYDRO1K, with a spatial resolution of 1 × 1 km derived from the GTOPO30 digital elevation dataset. HYDRO1K was developed by the U.S. Geological Survey in cooperation with UNEP/GRID (HYDRO1K, 2002), with additional funding from the Brazilian Water Resources Secretariat and the Food and Agriculture Organization/Inland Water Resources and Aquaculture Service. It provides users on a continent by continent basis with a low-resolution digital elevation model, along with ancillary datasets for use in continental and regional applications. The following prod-

BOX 5-7
Three Possible Methods for Estimating Population Using Remotely Sensed Data

Dwelling Unit Approach: The most accurate remote sensing-assisted method of estimating the population is to count individual dwelling units (see Lo, 1995; Haack et al., 1997; Jensen, 2000). If remotely sensed imagery with a spatial resolution of 0.25 to 5 m is available, this methodology works reasonably well for local censuses. Unfortunately, it is not suitable for a regional or national census of population because it is too time-consuming and costly (Sutton et al., 1997). In fact, Broome (U.S. Bureau of the Census, personal communication, 1998) has suggested that this method requires so much in situ data to calibrate the remote sensor data that it can become operationally impractical. Furthermore, the costs of acquiring the needed data are prohibitive. Therefore, other methods have been developed.

Total Urbanized Area Approach: Urban population correlates fairly consistently with total urbanized area at sub-national scales (Olorunfemi, 1984). Sutton et al. (1997) used Defense Meteorological Satellite Program Operational Linescan System (DMSP-OLS) nighttime 1 × 1 km imagery to inventory urban extent for the entire United States, with promising results at the state and county level. Unfortunately, this method "may underestimate the population density of urban centers and overestimate the population density of suburban areas."

Land Use Approach: This approach assumes that land use in an urban area is closely correlated with population density. First, a value is established for the population density for each land use by field survey or census data (e.g., multiple-family residential land use may contain 10 persons per pixel when using 30 × 30 m Landsat TM data, whereas rural forested areas might have only 0.20 persons per pixel). Next, by measuring the total area for each land-use category, the total population for that category is estimated. Finally, adding together the estimated totals for each land-use category provides the total population projection (Lo, 1995).

BOX 5-8
GPS Helps Resolve a Border Dispute

In 2000 the Tanzanian and Ugandan governments used GPS to resolve their border dispute that began in 1978. The Ugandan government gave up its territorial claim to a strip of land on the common border with Tanzania after GPS measurements proved that the pillars demarcating the two countries were 300 m inside Tanzania. As a result of the survey Uganda has moved the pillars of the border, a 100-kilometer straight line, to new positions inside Uganda.

SOURCE: People's Daily (2000).

ucts are available for Africa: streams,[24] drainage basins,[25] flow direction, flow accumulation, elevation, compound topographic index, slope, and aspect (HYDRO1K, 2002).

A valuable hydrologic product for application to Agenda 21 issues could be derived from the Shuttle Radar Topography Mission with almost global 90 × 90 m (perhaps 30 × 30 m) spatial resolution. This derivative product would have applications at the sub-regional level where HYDROlK currently is inapplicable. **Serious consideration should be given by the USGS to modeling the Shuttle Radar Topography Mission-derived 30 × 30 m digital elevation data to produce the most accurate, affordable hydrologic network database with global coverage.**

Government Units

Boundaries of government units, such as national borders, can be contentious, yet they are not always well documented (e.g., Box 5-8). One method for determining a boundary is to collect in situ GPS measurements. Alternatively, remotely sensed ortho-images (e.g., the global GeoCover-ortho dataset) can be used if the political boundary lies along a feature such as a river centerline, mountain ridge, or shoreline. For greater precision, boundaries can be delineated from very high spatial resolution imagery (e.g., less than 1 × 1 m) obtained in stereo.[26] However, remotely sensed imagery is inappropriate when the boundary is not visible in the imagery. Consequently, boundary-mapping endeavors often rely on both in situ surveying and remote-sensing.

The "Global GIS Database" (Hearn et al., 2001) contains boundary information with global coverage. Developed by the USGS through an agreement with ESRI, the database is of sufficient spatial resolution (1 × 1 km: or a scale of 1:1,000,000) for use as a regional reference and analytical tool but has limited value at the local level. NIMA's Operational Navigation Chart series is the primary data source for the "political boundaries" data set in the database. This series is the largest-scale (highest spatial resolution) unclassified map series that provides global coverage of features such as political boundaries. The "gazetteer" in the Global GIS Database, which grows by roughly 20,000 features monthly, currently contains 3.5 million geographic names from NIMA's database of foreign geographic feature names. The database is read through free ArcView Data Publisher software, and is available by region for U.S.$10 on CD-ROMs or on a single DVD.

Transportation Networks

Transportation networks include roads, railways, and pipelines. In addition to improving basic services (e.g., delivery and collection) there are many benefits to mapping transportation networks (Table 5-4). Geographic information from GPS and remote-sensing technology is useful for managing transportation networks (DOT, 2001) (Table 5-4). As developing countries move toward adopting these kinds of technologies, they can use other sources of information, such as the Global GIS Database (for regional applications) and legacy maps. However, these sources have limited value for the applications in Table 5-4 because of their coarse spatial resolution. Additionally, transportation networks can rapidly evolve, particularly in urban areas, and up-to-date, accurate information for effective transportation network management is needed (e.g., Table 5-4).

Cadastral Information for Land Administration

A *cadastre* is a map accompanied by a register showing the ownership or possession of individual units of land that is used to facilitate efficient land administration and expedite land market transactions. Among its many applications to sustainable development (Table 2-5), a cadastre is critical for combating poverty, integrated environment and development decision-making, and sound management of solid, hazardous, and other waste.

In Africa, land often is owned communally. Ownership rights tend to be based on communal units of lineages and extended families. However, individual members have use rights or "usufruct" to the land they currently farm. The concept of individual land ownership, especially in urban areas, was one of the consequences of the colonial integration of African countries into the global free-market system. Ownership not only promotes individual care and concern about land but also enhances its value. Owning land provides individuals with economic assets that can be traded in land markets, used as collateral to raise credit or as security for vari-

[24]The larger perennial rivers and smaller rivers and streams were derived from the Digital Chart of the World (USGS, 2002a).

[25]Derived using vector stream networks along with a data set on flow direction (see Verdin and Jenson, 1996, for more information).

[26]A pair of overlapping images is obtained that allows the analyst to view the pair of images stereoscopically in three dimensions.

TABLE 5-4 Data Requirements for Application of Geographic Data to Transportation Network Management Challenges

Application	Explanation	Benefits	Data Requirements
Environmental assessment, integration, and streamlining	The growth of transportation networks associated with urban growth and sprawl generates such environmental impacts as deforestation, impact on local and regional hydrology, and accentuation of such land-atmosphere factors as the urban heat island effect.	Mitigation of the impacts of growing transportation networks.	Medium to coarse spatial resolution remote sensor data (e.g., 5 to 20 m resolution such as SPOT or Landsat Thematic Mapper) with moderate 16-day temporal resolution requirements.
Hazards, safety, and disaster management	Monitoring transportation infrastructure to maximize public safety during emergencies.	Minimize failures of transportation networks, and carefully and rapidly move people away from an area in an emergency.	Medium resolution remote sensor data (e.g., Landsat Thematic Mapper or SPOT imagery) with very high temporal resolution requirements, often less than a day.
Traffic surveillance, monitoring, and management	Managing the flow of people and goods between geographically separated locations. Relevant information includes vehicle traffic volumes, classifications, speeds, and truck weights.	Improved estimates in forecasting traffic flows (reduced congestion, better network design).	Airborne and satellite-based sensors are most applicable. Extremely high temporal (e.g., minutes to hours) and high spatial resolution data (usually $\leq$ 1 m) are needed.
Transportation infrastructure management	Maintenance, operation, and renewal of such assets as pavement, bridges, pipelines, rail lines, harbors, and airports. Historically, infrastructure management was conducted in situ on an asset-by-asset basis.	Information on the location and condition of these assets is critical to effective management.	Extremely high spatial resolution remote sensor data ($\leq$ 1 m) and in certain instances hyperspectral data (hundreds of bands).

SOURCE: DOT (2001)

ous forms of economic improvements. Because individual land ownership is nonexistent in large parts of rural Africa, except in eastern and southern Africa, challenges remain for rural Africans to obtain credit from lending institutions in their bid to improve quality of life.

Several capacities have facilitated privatization and ownership of land. These include the capacity to survey individual plots, identify their boundaries, determine their ownership or possession, formally register or informally recognize their status, and, where the system has developed in eastern and southern Africa, accord them formal titles. As a result of these capacities it is possible to register and monitor subsequent transactions in the plots of land over time, enabling effective cadastration of land or the development of a land information system of ownership and possession.

Typically, a cadastre is produced at a scale of 1:10,000 or larger. In certain instances high spatial resolution aerial photography or other remotely sensed data reveals boundaries associated with long-held communal property boundaries. However, these data are expensive. Consequently, the production of cadastres has low priority for most African countries and donor agencies, even when there are clear benefits. Although the production of cadastres has been stymied by their cost, GPS in concert with GIS is cheaper than traditional surveying techniques and should facilitate production of cadastres.

Notwithstanding these technological innovations, it needs to be stressed that cadastration in any country remains a complex, demanding, if essential exercise. It should be undertaken community by community so as to expedite the adjudication of disputed claims of ownership.

Continued development of cadastres could facilitate land management and administration, promote greater efficiency in the operation of land markets both in urban and rural areas, strengthen the operations of free-market economies, and reinforce the ability of governments to initiate and sustain land and agrarian reforms (e.g., de Soto, 2000). Over time, cadastres could play an indirect role in poverty reduction, especially through enhancing access to credit facilities for many people beyond the limits usually provided by the now popular microcredit schemes. Additionally, by inducing a strong proprietary interest in land, the process of cadastration can help to deepen environmentally sensitive attitudes to development. Last, the provision of cadastres facilitates the supply of socio-economic information for effective settlement management. Many cities and towns have no maps showing the network of streets and roads. Many streets and roads have no names and many houses on the roads have no numbers (ECA, 2001). In short, many Africans have no easily located residential addresses to facilitate their effective participation in social and economic transactions. These inadequacies have been one reason why the systematic delivery, management, expansion, and improvement of services to all segments of the population, the effective collection of taxes and rates, and the cost recovery for utilities and services have been difficult to implement in urban areas (ECA, 2001). **Because of the potential of cadastres to address Agenda 21 issues, including poverty reduction and land resource management, the U.S. government (USAID and USGS) should assist African countries in developing cadastres.**

SUMMARY

Framework geographic data that underlie many Agenda 21 issues are accessible at varying degrees to users in developing countries. Cost and availability are two factors that influence access, in addition to telecommunications infrastructures (Chapter 4). Some framework foundation data are affordable and available. These are geodetic control (which can be obtained using a GPS receiver) and ortho-imagery. Topographic framework foundation data will be affordable and available in the near future with better than 90×90 m spatial resolution, and there is potential for an equally high spatial resolution hydrologic network database. Currently, affordable global datasets on transportation and hydrologic networks are available but are of insufficient resolution for most applications. Conversely, essential framework data, such as human population distribution, government units, and cadastre, are currently unaffordable to many African countries, but could become available if financial and technical assistance is provided. Meanwhile, African countries have no other choice but to use legacy datasets that are out of date for such applications as managing transportation networks. Nonetheless, these datasets are invaluable as baselines for detecting change.

REFERENCES

Ahmed Saleck ben Mohamed Lemine ibn Bouh. 2001. Repertoire de Mohamed ben Tolba el Jacobi de Chinguetti de Mauritanie 1774-1856 avec Explication de Mohamed Abdalah ben Shbih ben Bouh. Nouakchott, Mauritania.

ASPRS (America Society of Photogrammetry and Remote Sensing). 2002. Grids and Datums; The Portuguese Republic. Photogrammetric Engineering and Remote Sensing 68(4):305-307.

Barnes, S. 2002. GNSS evolves. Geospatial Solutions 12(5):10-11.

British Museum. 2002. The British Museum Department of Libraries and Archives. Available at http://www.british-museum.ac.uk/libraries/index.html. Accessed on July 11, 2002.

CIESIN (Center for International Earth Science Information Network). 2002. Population density, 1995. Available at <http://sedac.ciesin.org/plue/gpw/main.html>. Accessed April 23, 2002.

De Soto, H. 2000. The Mystery of Capital: Why Capitalism Triumphs in the West and Fails Everywhere Else. New York: Basic Books..

DEVECOL (Development Ecology). 2002. Available at <http://www.devecol.org>. Accessed May 31, 2002.

Diop, L. 2001. Organization and Financing of Population Censuses in Sub-Saharan Africa: Problems and Prospects. In Symposium on Global Review of 2000 Round of Population and Housing Censuses: Mid-Decade Assessment and Future Prospects. New York: United Nations Secretariat.

Dobson, J. E., E. A. Bright, P. R. Coleman, R. C. Durfee, and B. A. Worley. 2000. A Global Population Database for Estimating Population at Risk, Photogrammetric Engineering and Remote Sensing, 66(7): 849-858.

Donaint, P. 1975 Les Cadres Géographiques à Travers les Langues du Niger—Contribution à la Pedagogie de l'Étude du Milieu. Etudes Nigeriennes No 37. Niamey: Institut Nigerien de Recherches en Science Humaines.

DOT (Department of Transportation). 2001. Commercial Remote Sensing Technologies Application to Transportation. Washington, DC: U.S. Department of Transportation and National Aeronautics and Space Administration.

Earth Satellite. 2002. GeoCover-Ortho Database, Rockville, MD: Earth Satellite, Inc. Available at <http://www.earthsat.com>. Accessed June 3, 2002.

ECA (U. N. Economic Commission for Africa). 2001. The Future Orientation of Geoinformation Activities in Africa: Synthesis. Available at <http://www.uneca.org/eca_resources/Conference_Reports_and_Other_Documents/disd/codi/docs/doc3EN.pdf>. Accessed May 15, 2002.

Elvidge, C. D., H. W. Kroehl, E. A. Kihn, K. E. Baugh, E. R. Davis, W. M. Hao. 1996 Algorithm for the Retrieval of Fire Pixels from DMSP Operational Linescan System. Pp. 73-85 in Global Biomass Burning, J. S. Levine, ed. Cambridge, MA: MIT Press.

Elvidge, C. D., K. E. Baugh, E. A. Kihn, H. W. Kroehl, and E. R. Davis. 1997. Mapping City Lights with Nighttime Data from the DMSP Operational Linescan System. Photogrammetric Engineering and Remote Sensing 63(6):727-734.

Elvidge, C. D., K. E. Baugh, V. R. Hobson, E. A. Kihn, and H. W. Kroehl. 2002. Detection of Fires and Power Outages at Night Using DMSP-OLS Data. Available at <dmsp@ngdc.noaa.gov>. Accessed Jne 3, 2002.

ESRI (Environmental Systems Research Institute). 2002. Saving the elephants with GIS and GPS. Available at <http://www.esri.com/news/arcnews/summer00articles/ savingelephants.html>. Accessed May 7, 2002.

Gesch, D. B., K. L. Verdin, and S. K. Greenlee. 1999. New Land Surface Digital Elevation Model Covers the Earth. Eos Transactions, American Geophysical Union 80(6):69-70.

GTOPO30 (Global Topography at 30 arc second spatial resolution). 2002. GTOPO30. Sioux Falls: U.S. Geological Survey EROS Data Center. Available at <http://edcdaac.usgs.gov/gtopo30/gtopo30.html>. Accessed June 3, 2002.

Haack, B., S. Guptill, R. Holz, S. Jampoler, J. Jensen, and R. Welch. 1997. Chapter 15: Urban Analysis and Planning. Pp 517-553 in Manual of Photographic Interpretation. Bethesda: American Society for Photogrammetry and Remote Sensing.

Hearn, P., Jr., T. Hare, P. Schruben, D. Sherrill, C. Lamar, and P. Tsushima. 2001. Global GIS Database: Digital Atlas of Africa, U.S. Geological Survey Digital Data Atlas Series DDS 62-B. Washington DC: U.S. Geological Survey, CD.

HYDRO1K. 2002. Sioux Falls: U.S. Geological Survey EROS Data Center. Available at <http://edcdaac.usgs.gov/gtopo30/hydro/>. Accessed May 31, 2002.

IGN (Institut Géographique National). 2002. Available at <http://www.ign.fr/>. Accessed July 11, 2002.

Jensen, J. R. 1995. Issues Involving the Creation of Digital Elevation Models and Terrain Corrected Orthoimagery Using Soft-Copy Photogrammetry. Geocarto International: A Multidisciplinary Journal of Remote Sensing 10(1):1-17.

Jensen, J. R. 2000. Remote Sensing of the Environment: An Earth Resource Perspective. Saddle River: Prentice-Hall.

Kobrick, M. 2002. Pictures from the Real Edge: NASA Posts U.S. Topography Data. Pasadena: NASA Jet Propulsion Laboratory Media Relations Office. Available at <http://www.jpl.nasa.gov>. Accessed June 3, 2002.

LandScan. 2000. Landscan Global Population Database, Oakridge, TN: Oak Ridge National Laboratory. Available at <http://sedac.ciesin.columbia.edu/plue/gpw/ landscan>. Accessed June 3, 2002.

Leete, R. 2001. Population and Housing Censuses: A Funding Crisis? In Symposium on Global Review of 2000 Round of Population and Housing Censuses: Mid-Decade Assessment and Future Prospects. New York: United Nations Secretariat.

Liverman, D., E. F. Moran, R. R. Rindfuss, and P. C. Stern (eds). 1998. People and Pixels—Linking Remote Sensing and Social Science. Washington, DC: National Academy Press.

Lo, C. P. 1995. Automated Population and Dwelling Unit Estimation from High-Resolution Satellite Images: A GIS Approach. International Journal of Remote Sensing 16:17-34.

McCanna, R. L. 2002. Shuttle Radar Topography Mission (SRTM) Data. Briefing to the Committee on the Geographic Foundation of Agenda 21, Washington, DC: National Academy of Sciences.

Murray, J. A. (ed). 1993. Wild Africa—Three Centuries of Nature Writing from Africa. Oxford: Oxford University Press.

NASA (National Aeronautics and Space Administration). 2002. Available at <http://www.esad.ssc.nasa.gov/datapurchase/default.asp?sec=background>. Accessed: May 1, 2002.

NRC (National Research Council). 2001. Transforming Remote Sensing Data into Information and Applications. Washington, DC: National Academy Press.

Olorunfemi, J. F. 1984. Land Use and Population: a Linking Model. Photogrammetric Engineering and Remote Sensing 50:221-227.

Parkinson, B. W., 1994. GPS Eyewitness: The Early Years. GPS World, 5(9): 32-45.

People's Daily. 2000. Tanzania, Uganda Resolve Border Dispute. Available at: <http://english.peopledaily.com.cn/200012/18/eng20001218_58164.html>. Accessed on May 28, 2002.

Rizos, C. 2002. Introducing the Global Positioning System. Pp 77-94 in Manual of Geospatial Science and Technology, J. D. Bossler, J. R. Jensen, R. B. McMaster, and C. Rizos, eds. London: Taylor and Francis.

SRTM (Shuttle Radar Topography Mission), 2002a. Shuttle Radar Topography Mission. Available at http://www.jpl.nasa.gov/srtm/. Accessed July 11, 2002.

SRTM. 2002b. SRTM Fact Sheet: Seeing Earth's Surface in 3-D. Pasadena: NASA Jet Propulsion Laboratory. Available at <http:// www.jpl.nasa.gov/srtm/ factsheet_pub.html>.

Stewart, M. and C. Rizos. 2002. GPS projects: Some Planning Issues. Pp 162-182 in Manual of Geospatial Science and Technology. J. Bossler, J. Jenson, R. McMaster, and C. Rizos, eds. London: Taylor and Francis.

Sutton, P., D. Roberts, C. Elvidge, and H. Meij. 1997. A Comparison of Nighttime Satellite Imagery and Population Density for the Continental United States. Photogrammetric Engineering and Remote Sensing 63(11):1303-1313.

Taylor, D. R. F. 1997. Modern Cartography, Policy Issues and the Developing Nations. Pp. 185-213 in Policy Issues in Modern Cartography, D.R.F. Taylor, ed. Oxford: Pergamon.

Thrower, N.J. and J. R. Jensen. 1976. The Orthophoto and Orthophotomap: Characteristics, Development, and Aspects of Cartographic Communication. American Cartographer 3(1):39-56.

UN (United Nations). 1998. Principles and Recommendations for Population and Housing Censuses. New York: United Nations.

UNDP (United Nations Development Programme), 2001. World Population Prospects Population Database. Available at <http://esa.un.org/unpp/sources.html>. Accessed June 3, 2002.

UNFPA (U.N. Population Fund). Unpublished. Compendium of Census Activities in Sub-Saharan Africa. 2000 Round of Censuses. Compiled by the Africa Division, UNFPA.

USGS (U.S. Geological Survey). 1998. Earthshots: Satellite Images of Environmental Change. EROS Data Center. Available at <http://edcwww.cr.usgs.gov/earthshots/ slow/tableofcontents>. Accessed June 26, 2002.

USGS. 1999. The Global Positioning System, Fact Sheet 062-99. Washington, DC: U.S. Geological Survey.

USGS. 2002a. Digital Chart of the World–DOS, Sioux Falls, SD: EROS Data Center, U.S. Geological Survey. Available at: <http://edc.usgs.gov/glis/hyper/oldguides/dcw>. Accessed July 11, 2002.

USGS. 2002b. GTOPO30 Order Form. Available at <http://edcdaac.usgs.gov/gtopo30/ form.html>. Accessed July 11, 2002.

USGS, 2002c. Landsat Data Continuity Mission Homepage. Available at <http://ldcm.usgs.gov/>. Accessed July 11, 2002.

Verdin, K. L., and S. K. Jenson. 1996. Development of Continental Scale DEMs and Extraction of Hydrographic Features, Proceedings, National Center for Geographic Information and Analysis International Conference/Workshop on Integrating GIS and Environmental Modeling. Santa Fe, NM, January 21-25, CD, 1 disc.

Welch, R. 1980. Monitoring Urban Population and Energy Utilization Patterns from Satellite Data. Remote Sensing of Environment 9:1-9.

ANNEX 5

ANNEX BOX 5-1
Components of the NAVSTAR Global Positioning System

NAVSTAR consists of three components: the *space segment* (consisting of the satellites and the transmitted signals); the *control segment* (which tracks the satellites and gives them instructions); and the *user segment* (where GPS data are turned into spatial information for use in many applications).

Space Segment

This segment consists of 24 satellites and the signals that are broadcast from them. This information allows users to determine their position, velocity, and time. Only a certain number of satellites above the observer's horizon are "visible" to a GPS unit at any moment in time. This has a significant impact on the precision of the positioning measurement. The greater the number of satellites above the observer's horizon, the higher the quality of the measurement. Each GPS satellite transmits unique navigational signals centered on two *L*-band frequencies (carrier waves) plus ranging codes modulated on the carrier waves, and a navigation message. The ranging codes are used to compute the signal transit time from the satellite transmitter to the receiver. Multiplying this value by the speed of light yields the range from the receiver to the satellite. When enough of these measurements are obtained from the various satellites above the observer's horizon, it is possible to triangulate on the position of the observer.

The Control Segment

The U.S. Department of Defense operates five ground stations: in Hawaii, Colorado, Ascension Island, Diego Garcia, and Kwajalein. These stations communicate with the satellites, adjust their orbits, and maintain the constellation of satellites and their functions.

The User Segment

The user segment results in the conversion of range information measured by the GPS unit into useable positioning information. The accuracy of these coordinates is a function of several parameters. For example, the observer may select absolute (single-point) positioning or relative positioning. Generally, the relative positioning system achieves more accurate results. *Absolute positioning* refers to use of a single GPS system in communication with the constellation of satellites. Conversely, *relative positioning* measurements are normally obtained using two GPS units on the ground. For relative positioning, one of the GPS units functions as the base station while the other GPS unit is carried by the user to locations of interest. The relative position is the difference between the two positions (in the global system), expressed in a local reference system with the origin at the base station (Rizos, 2002). The accuracy of relative positioning measurements is a function of the length of the baseline between GPS receivers, the length of time and the number of observations obtained at a specific location by the GPS unit, the number and position of available satellites, and the amount and quality of the post-processing of raw data (Stewart and Rizos, 2002). Accuracies in the centimeter range are attainable for geodetic and general way-finding under certain conditions. The accuracy of the elevation measurement at each location is about two to three times poorer than the horizontal positioning accuracy. Therefore, it is important to achieve as accurate a horizontal positioning as possible.

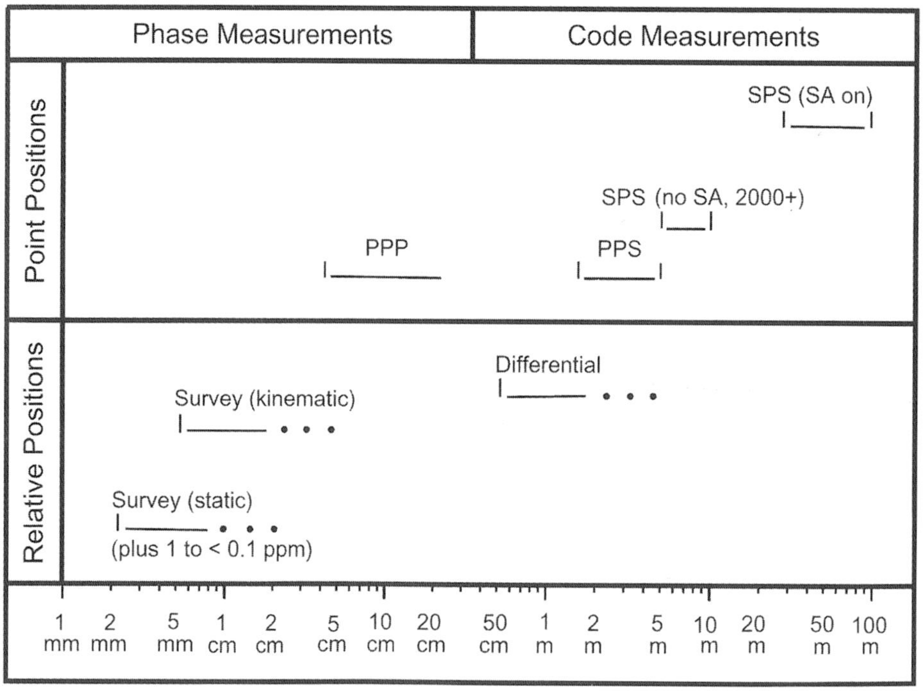

FIGURE 5-A1 Global Positioning System accuracy and positioning modes (adapted from Rizos [2002]). The top half of the illustration refers to single point positioning measurements made with a single GPS unit communicating with a number of satellites. The lower half refers to the relative positioning mode where the position of the receiver of interest (i.e., the location of the user) is derived relative to a base station receiver with known absolute coordinates.

ANNEX BOX 5-2
Remote-Sensing Resolution Considerations

Spatial resolution is a measure of the smallest angular or linear separation between two objects that can be resolved by the remote-sensing system. The spatial resolution of aerial photography may be measured by (1) placing carefully calibrated, parallel black-and-white lines on tarps that are placed in the field, (2) obtaining aerial photography of the study area, and (3) analyzing the photography and computing the number of resolvable line pairs per millimeter in the photography. For electronic remote-sensing systems, the nominal spatial resolution is the dimension in meters (or feet) of the ground-projected instantaneous-field-of-view (Jensen, 2000). For example, the IKONOS panchromatic band has a nominal spatial resolution of 1 × 1 m and the Landsat Thematic Mapper 5 has a nominal spatial resolution of 30 × 30 m for six of its bands. Generally, the smaller the spatial resolution, the greater the resolving power of the sensor system. Simulated examples of different spatial resolution remote sensor data are shown in Figure 5-A2.

Temporal resolution refers to how often remotely sensed data are acquired over a particular geographic area. For example, the temporal resolution of the polar-orbiting Landsat 4 and 5 Thematic Mapper remote-sensing systems has been 16 days. If a remote-sensing system can be pointed off-nadir (i.e., it does not have to look straight down), then it is possible to obtain much higher temporal resolution (e.g., the pointable Space Imaging, Inc., IKONOS and DigitalGlobe Quickbird can obtain imagery every few days, depending upon the latitude of the area of interest). Some satellites such as the Geostationary Operational Environmental Satellites (GOES) are located in an orbit a certain distance above a particular point on the ground. Such remote-sensing systems often have very high temporal resolution (e.g., they obtain imagery every one-half hour) that facilitates the tracking of tornadoes, frontal systems, and hurricanes.

Spectral resolution is the number and dimension of specific wavelength intervals in the electromagnetic spectrum to which a remote-sensing instrument is sensitive. Remote-sensing systems may be configured to collect data in just a single band of the electromagnetic spectrum. For example, a digital frame camera band 4 near-infrared image is displayed in Figure 5-A3a. *Multispectral* remote-sensing takes place when energy is recorded in multiple bands of the electromagnetic spectrum. For example, the ADAR 5500 usually acquires four multispectral bands of imagery during a mission as shown in Figure 5-A3b. The bandwidths are shown in Figure 5-A3c: band 1 = 450-515 nm; band 2 = 525-605 nm; band 3 = 640-690 nm; band 4 = 750-900 nm. A *hyperspectral* remote-sensing instrument acquires data in hundreds of spectral bands. For example, the Airborne Visible and Infrared Imaging Spectrometer (AVIRIS) has 224 bands in the region from 0.4-2.5 μm spaced just 10 nm apart. *Ultraspectral* remote-sensing involves data collection in many hundreds of bands. Careful selection of the spectral bands improves the probability that a feature or biophysical parameter (e.g., biomass, turbidity) will be detected, identified, and measured accurately.

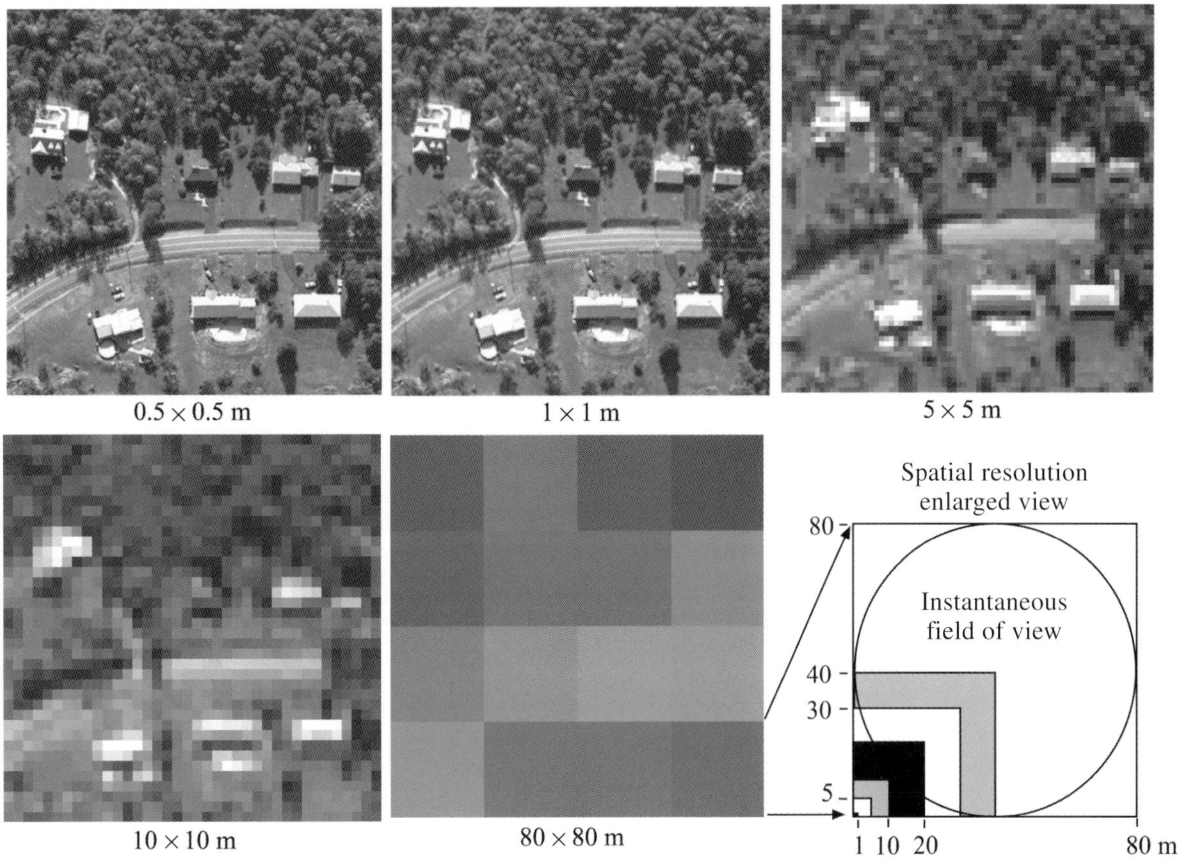

FIGURE 5-A2 Remote sensor data of Mechanicsville, N.Y. on June 1, 1998, at a nominal spatial resolution of 0.3 × 0.3 m (approximately 1 × 1 ft.) using a digital camera (courtesy of E-ConAgra.com). The original data were re-sampled to derive the imagery with the simulated spatial resolutions shown.

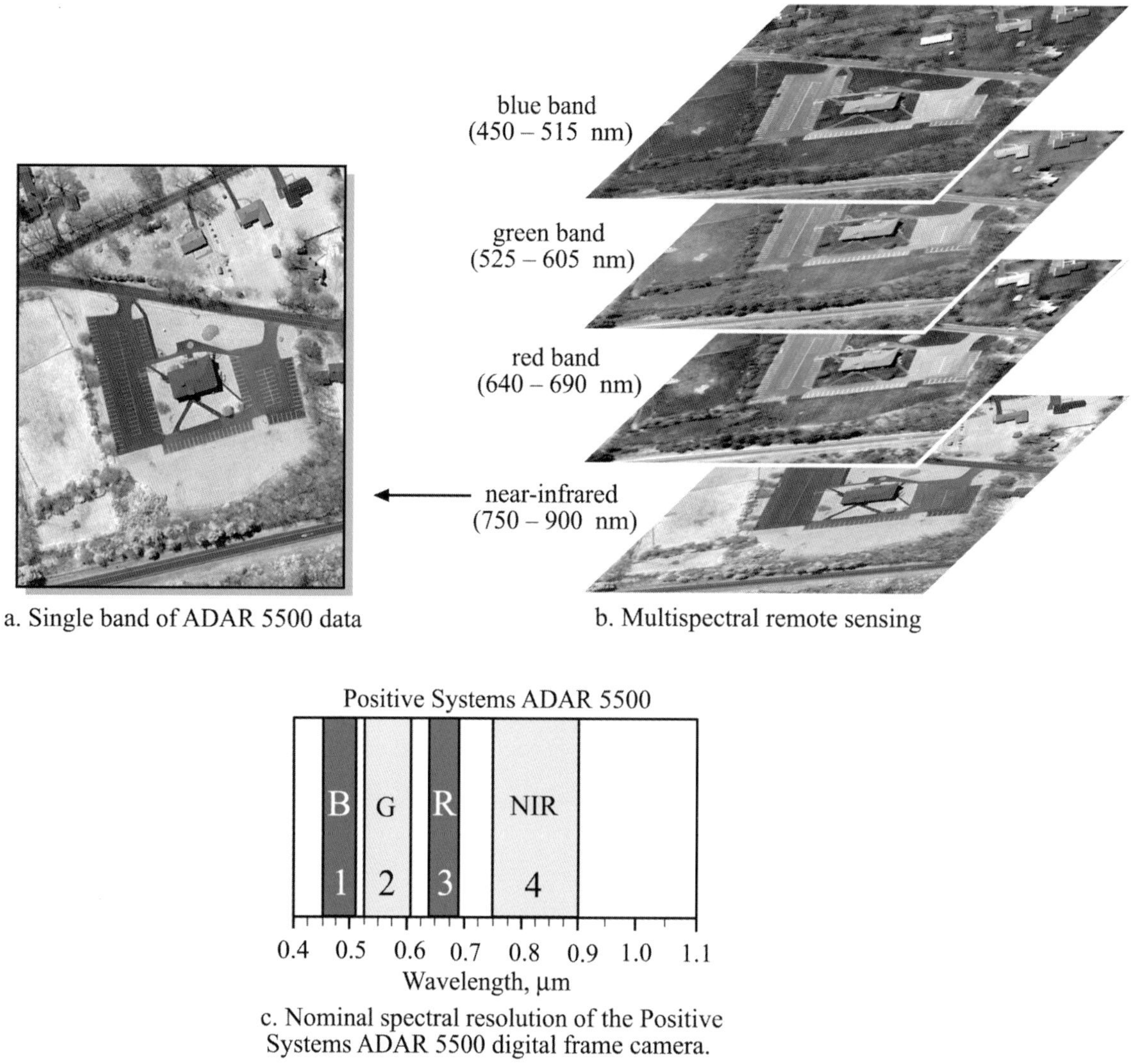

FIGURE 5-A3 An example of single band and multispectral remote-sensing using a digital frame camera. The data were obtained at a spatial resolution of 1 × 1 ft. (courtesy of Positive Systems, Inc.).

ANNEX BOX 5-3
Landsat 4 and 5 Thematic Mapper
Landsat 7 Enhanced Thematic Mapper Plus
Landsat Data Continuity Mission

NASA launched its first Landsat satellite in July 1972, (originally called ERTS, Earth Resources Technology Satellite). This satellite collected data at 79 × 79 m spatial resolution in four bands. Since 1972, five Landsat satellites have been placed into orbit, two of which were thematic mappers that obtained data at a spatial resolution of 30 × 30 m in seven bands.

NASA's Landsat Thematic Mapper 4 and 5 were launched on July 16, 1982, and March 1, 1984, respectively. Landsat 5 is still functioning. The Earth Observation Satellite Company (now Space Imaging) obtained control of the Landsat 4 and 5 satellites in September 1985. Thematic Mapper 4 and 5 satellites were placed in Sun-synchronous polar orbits at an altitude of 705 km. The TM sensors are based on optical-mechanical scanning technology. They collect data in a swath 185 km wide but cannot view off-nadir. The revisit cycle (temporal resolution) is every 16 days. The TM bands were selected to make maximum use of the dominant factors controlling leaf reflectance, such as leaf pigmentation, leaf and canopy structure, and moisture content. The spectral and spatial resolution characteristics are shown below.

Wavelength (μm)		Spatial Resolution (m)	Sensor System
Band 1	0.45-0.52 [0.450–0.515]	30 × 30	TM 4, 5, [ETM+ 7]
Band 2	0.52-0.60 [0.525–0.605]	30 × 30	TM 4, 5, [ETM+ 7]
Band 3	0.63-0.69	30 × 30	TM 4, 5, ETM+ 7
Band 4	0.76-0.90 [0.750–0.900]	30 × 30	TM 4, 5, [ETM+ 7]
Band 5	1.55-1.75	30 × 30	TM 4, 5, ETM+ 7
Band 6	10.40-12.50	120 × 120 [60 × 60]	TM 4, 5, [ETM+ 7]
Band 7	2.08-2.35	30 × 30	TM 4, 5, ETM+ 7
Band 8	0.52-0.90	15 × 15	ETM+ 7

Much of the Landsat TM 4 and 5 data are available at reasonable rates from the EROS Data Center in Sioux Falls, S. Dak., and from Space Imaging, Inc. Landsat 6 failed to achieve orbit in 1993.

NASA's Landsat 7 Enhanced Thematic Mapper Plus (ETM+) was launched on April 15, 1999. The ETM+ was developed to be in harmony with the relatively coarse spatial resolution sensors onboard NASA's Earth Observing System Terra satellite as it provides a unique suite of relatively high spatial resolution observations. In addition, Landsat 7 was designed to (a) maintain data continuity by providing data that are consistent in terms of geometry, spatial resolution, coverage, and spectral characteristics with previous Landsat data, (b) periodically refresh a global archive of cloud-free, sun-lit landmass imagery, and (c) continue to make Landsat-type data available to international users at the cost of fulfilling user requests.

Landsat 7 is based on the same scanner technology as Landsat 4 and 5. However, the ETM+ has significantly better radiometric calibration than its predecessors. ETM+ bands 1-5 and 7 are almost identical to those found on Landsat 4 and 5 and have the same 30 × 30 m spatial resolution with two exceptions. The thermal infrared band 6 (10.4–12.5 μm) has 60 × 60 m spatial resolution. There is a new 15 × 15 m panchromatic band (0.52–0.90 μm). Landsat 7 data are controlled by the U.S. Geological Survey. The products cost $475 (level 0R) or $600 (level 1R and 1G). There are no restrictions on the use, reprocessing, or redistribution of Landsat 7 data purchased from the U. S. government.

Landsat Data Continuity Mission. The Land Remote Sensing Policy Act of 1992 directs Landsat program management (NASA and the USGS) to assess options for a data system to succeed the orbiting Landsat 7. This is called the Landsat Data Continuity Mission. Currently, two vendors (Resource-21 and DigitalGlobe) have been funded to provide formulation contracts. NASA expects to finalize the award for the Landsat Data Continuity Mission in mid-2003, with data delivery to the government in 2005 (USGS, 2002c)

ANNEX BOX 5-4
Processing NASA's Global GeoCover-Ortho Imagery

The Global GeoCover-Ortho database consists of Landsat Thematic Mapper images obtained circa 1990 that have the following characteristics:

- Collected within 3 years of 1990 target date;
- < 20 percent cloud cover;
- Within the uppermost 10 percent in terms of quality rating; and
- Phenology (plant development cycle) at or just past peak "greenness."

The images were obtained from the Landsat foreign ground receiving stations,[1] as well as from the U.S. Geological Survey's EROS Data Center. The complete GeoCover-Ortho database includes 7,100 Landsat TM images. The positional accuracy of the Global GeoCover ortho-rectified imagery is better than 50 m.[2] This positional accuracy is superior to the vast majority of the world's 1:200,000 scale maps. The following description explains how this was achieved.

To transform a Landsat TM image to a standard map projection scientists typically manipulate (rectify) it by selecting many ground control points and then applying mathematical techniques. This is a very time-consuming and expensive process that can result in inconsistencies at the edges of the image when adjacent TM images are processed independently. Consequently, it is difficult to process a large number of images using this approach (Jensen, 1995). EarthSat had the task of orthorectifying thousands of Landsat TM (circa 1990) and Multispectral Scanner (from the mid 1970s) images for the entire land surface of Earth in conjunction with its NASA contract (Earth Satellite, 2002). To do this accurately and efficiently a photogrammetric approach to the problem was used.

A proprietary photogrammetric methodology was used to process groups of up to 400 Landsat images of which <10 percent contained ground control points against which they could be corrected. To create an accurate spatial relationship between the raw Landsat imagery and Earth's surface the approach combined the world's best horizontal and vertical ground control points with precisely measured tie points between adjacent images. Wherever possible, Earthsat used U.S. government digital elevation data with a spacing of 90 m for the vertical control. These were backed up by 900-m spaced data from the GTOPO30 data set when the 90-m data were not available.

[1]These ground stations receive data from passing Landsat satellites.

[2]Root mean square accuracy in a Universal Transverse Mercator projection based on the WGS 84 datum (Earth Satellite, 2002).

ANNEX BOX 5-5
Processing Shuttle Radar Topography Mission Data for Public Release

SRTM processing is occurring in two parallel tracks:

- ***Track 1:*** Continent-scale datasets are being processed by NASA on a continent-by-continent basis (Kobrick, 2002). North America was processed first and delivered in July 2002. As Africa and other continents are completed, the data will be refined by NIMA and then distributed to the public through the USGS EROS Data Center. Africa is projected for completion in late 2003 to mid-2004. Access to SRTM-derived data by developing countries is through bilateral agreements (McCanna, 2002).

- ***Track 2:*** Smaller datasets (covering a number of 1 × 1 degree latitude and longitude blocks) are being processed for sites of scientific interest to NASA-sponsored researchers. Each dataset consists of unedited digital elevation maps, images, and ancillary data. Upon completion they will be made available to the scientific community and the public. Several of these datasets have already been prepared for various parts of Africa (e.g., Figures 5-8 and 5-9).

ANNEX BOX 5-6
Characteristics of the Defense Meteorological Satellite Program

Each DMSP satellite crosses any point on Earth up to twice each day (equatorial crossing times are 0536 and 1052 local time). Because there are multiple satellites, nearly complete global coverage of clouds and features on Earth's surface is achieved every six hours.

The current generation of Operational Linescan System sensors began flying in 1976 and is expected to continue flying until approximately 2008, when it will be officially merged with the National Polar-orbiting Operational Environmental Satellite System (NPOESS). Since March 1992, data have been sent daily to the U.S. National Geophysical Data Center for archiving. DMSP data are processed within five days of receipt. The data are available worldwide for the cost of materials and mailing. The archived data set consists of low spatial resolution (2.7 × 2.7 km) imagery with global coverage and high spatial resolution (0.55 × 0.55 km) imagery with regional coverage (along a 3,000 km scan).

DMSP satellites are in a near-polar, Sun-synchronous orbit at an altitude of 830 km above Earth. Each orbit lasts approximately 101 minutes. The Operational Linescan System consists of two telescopes and a photo multiplier tube. Each sensor sweeps back and forth on a line on the Earth's surface in a whiskbroom motion. The photo multiplier tube is used to detect nighttime lights in the range from 0.47 to 0.95 μm. One of the telescopes detects emissions from the Sun or the Moon that are *reflected* off clouds and features on the surface of Earth (in the range from 0.40 to 1.10 μm). The other telescope records thermal *emissions* by Earth's surface and atmosphere (in the range from 10.0-13.4 μm).

ANNEX TABLE 5-1 Characteristics of Global GeoCover-Ortho Image Products (Earth Satellite Corporation)

Image Type	Characteristics
Individual images	Images cover an area of approximately 170 × 170 km, with a spatial resolution of 28.5 m. The complete GeoCover-Ortho database includes 7,100 orthorectified Landsat TM images. Images consist of all seven Landsat spectral bands in a Universal Transverse Mercator map projection with a geodetic accuracy of better than 50 m (root mean square accuracy). Images can be obtained with unique projections, datums, and band combinations.
Mosaics	Each mosaic is a group of juxtaposed Landsat images in a single, seamless digital image. Mosaics are three-band color composite products based on Landsat TM spectral bands 7,4,2. Mosaics may contain up to 15 Landsat images covering a 5-degree (north-south) by 6-degree (east-west) segment of a Universal Transverse Mercator grid. The mosaics have the same spatial resolution, geodetic accuracy, and map projection as the individual images.
Regional mosaics	These are large-area mosaics of 1,200+ Landsat images. Typically they comprise Landsat TM bands 7,4,2. The imagery has a spatial resolution commensurate with the scale of the final digital mosaic, but usually between 90 and 150 m. The map projections of regional mosaics vary and are determined by the conventions for the areas being covered. The following GeoCover-Ortho mosaics are available: Africa; northern Africa; Alaska; Australia; Central America; Middle East; northeast Asia; United States.

ANNEX TABLE 5-2 Census Dates for Countries in Africa from 1965 to 2004

	1970 Round 1965-1974	1980 Round 1975-1984	1990 Round 1985-1994	2000 Round 1995-2004
Algeria	1966	1977	1987	1998
Angola	1970	—	—	2002 P
Benin	—	1979 F	1992	2002 S
Botswana	1971	1981	1991	2001 S
Burkina Faso	—	1975 F	1985	1996
Burundi	—	1979 F	1990	—
Cameroon	—	1976 F	1987	2002 S
Cape Verde	1970	1980	1990	2000
Central African Republic	—	1975 F	1988	2002 S
Chad	—	—	1993 F	2004-2005
Comoros	1966	1980	1991	2002 S
Congo	1974 F	1984	—	1996: 2000A
Cote d'Ivoire	—	1975 F	1988	1998
Democratic Republic of the Congo	1970 A;1974 A	1984	—	—
Djibouti	1967; 1970-71 A	1983 F	—	2001 S
Egypt	1966 SC	1976	1986	1996
Equatorial Guinea	1971 A	1983	1994	2002 S
Eritrea	—	1984 by Ethiopia	—	—
Ethiopia	—	1984 F	1994	2004 S
Gabon	1970	1980	1993	2003 P
Gambia	1973	1983	1993	2003 P
Ghana	1970	1984	—	2000
Guinea	1967 A; 1972 A	1977 A; 1983 F	—	1996
Guinea Bissau	1970	1979	1991	2001 S
Kenya	1969	1979	1989	1999
Lesotho	1966	1976	1986	2001
Liberia	1974	1984	—	2003 S
Libya	1973	1984	—	1995
Madagascar	—	1975 F	1993	2003 S
Malawi	1966 F	1977	1987	1998
Mali	—	1976 F	1987	1998
Mauritania	—	1976-77 F	1988	2000-01 S
Mauritius	1972	1983	1990	2000
Morocco	1971	1982	1994	—
Mozambique	1970	1980	—	1997
Namibia	1970	1981	1991	2001
Niger	—	1977 F	1988	2001
Nigeria	—	—	1991	2004 S
Reunion	1974	1982	1990	1999
Rwanda	—	1978 F	1991	2002 S
Sao Tome and Principe	1970	1981	1991	2001 S
Senegal	—	1976 F	1988	2003 S
Seychelles	1971	1977	1987;1994	2004 P
Sierra Leone	1974	—	1985	2003 S
Somalia	—	1975 F	1987	—
South Africa	1970	1980	1985; 1991	1996; 2001
Sudan	1973	1983	1993	2003 P
Swaziland	1966	1976	1986	1997
Tanzania	1967	1978	1988	2002 S
Togo	1970 F	1981	—	2002 S
Tunisia	1966	1975	1994	2004 P
Uganda	1969; 1974	1980	1991	2002 S
Western Sahara	1970; 1974	1982 by Morocco	1994 by Morocco	—
Zambia	1969; 1974 SC	1980	1990	2000
Zimbabwe	1969	1982	1992	2002 S

— No census listed in this round.
A Administrative census.
S Scheduled; not yet taken or known if taken.
F First full modern census taken.
P Projected based on pattern of census dates.
SC Sample census.

SOURCE: U.S. Census Bureau; UNFPA (unpublished).

6

Geographic Data for Sustainable Development II: Other Thematic Data

INTRODUCTION

Drawing on examples of remotely sensed satellite data that are mostly low in cost, this chapter describes the sources, adequacy, and current applications of important thematic data types for monitoring and managing natural and human-made resources in Africa. These data types form the organizational framework for the chapter. First, the chapter addresses land-cover and land-use data (e.g., depicting agriculture, savannah, forest, settlements). Second, it examines biophysical data (e.g., rainfall, and data relating to the physical condition of vegetation). Finally, it describes data for managing human health (e.g., environmental data pertaining to vector habitats). These thematic data types supplement the framework foundation data (Chapter 5) that form the core of a country's geographic data needs for addressing Agenda 21 issues. Much of the technical information on data sources in Chapter 6 is found in Annex 6.

LAND COVER AND LAND USE

The pace, magnitude, and scale of human alterations of Earth's land surface are unprecedented in human history. Consequently, land-cover and land-use data are central to such Agenda 21 issues as combating deforestation, managing sustainable settlement growth, and protecting the quality and supply of water resources (Table 2-5). In light of the human impacts on the landscape, there is a need to establish baseline datasets against which changes in land cover and land use can be assessed. "Land cover" refers to the type of material present on the landscape (e.g., water, sand, crops, forest, wetland, human structures). "Land use" refers to what people do on the land surface (e.g., agriculture, commerce, settlement).

The International Geosphere-Biosphere Programme (IGBP) and the International Human Dimensions of Global Environmental Change Programme (IHDP) suggest that

> [o]ver the coming decades, the global effects of land use and cover change may be as significant, or more so, than those associated with potential climate change. Unlike climate change per se, land use and cover change are known and undisputed aspects of global environmental change. These changes and their impacts are with us now, ranging from potential climate warming to land degradation and biodiversity loss and from food production to spread of infectious diseases (IGBP-IHDP, 2002).

In addition to understanding changes that have already occurred, land-cover data are needed to generate scenarios of future modification of the Earth system (Lambin and Geist, 2001; Geist and Lambin, 2002).

Land-use and land-cover data can be obtained using in situ field measurements or remote-sensing technology. However, access to raw remotely sensed data alone is insufficient to feed decision-support systems. To extract useful thematic information such as land-cover maps from the raw imagery decision-makers must rely on intermediate steps involving scientific expertise, use of calibration data, and image-processing resources.

Different applications of land-use and land-cover information normally require that remotely sensed data be obtained at different spatial resolutions. For convenience the land-cover information is often grouped into four levels that can be associated with remotely sensed data acquired at different spatial resolutions (Anderson et al., 1976) (Figure 6-1). Level I nominal-scale land-cover information might identify an area as forested. Level II might make a further distinction between deciduous and coniferous forest. Level III might include information on particular species (e.g., acacia). Level IV might include sub-species information. The extremely high level of detail needed for land cover Levels III and IV is usually derived from high spatial resolution remote-sensor data such as that provided by large-scale aerial photography or certain commercial satellite remote-sensing systems. Information may be extracted using classical photo-

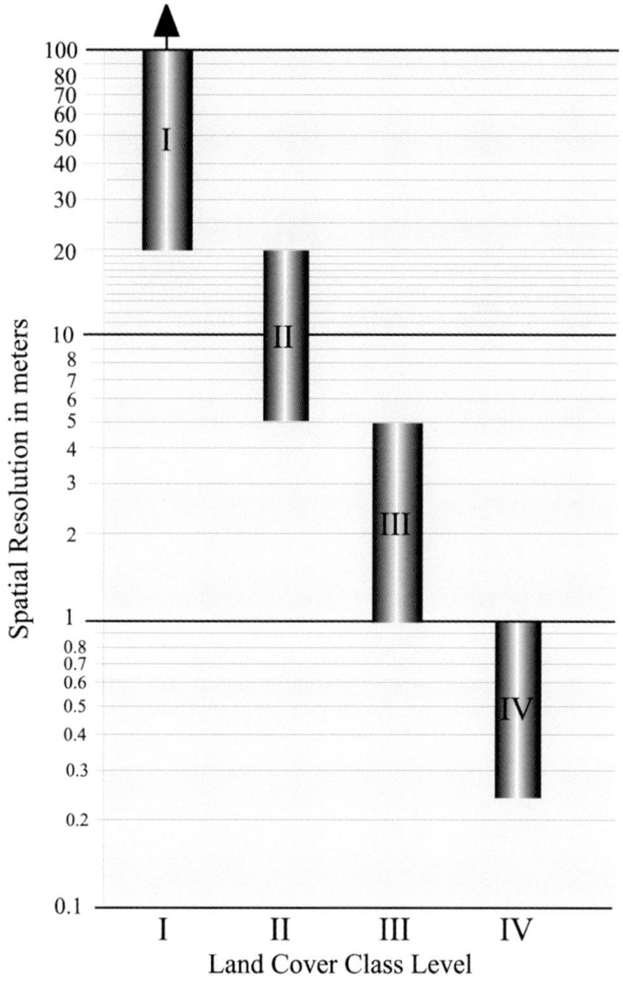

FIGURE 6-1 The relationship between U.S. Geological Survey land-cover and land-use classes and the required spatial resolution of the imagery. SOURCE: Pearson Education, Inc., adapted from Jensen (2000).

interpretation techniques applied to analog (hard-copy) imagery or digital image processing techniques applied to digital remote-sensor data (including digitized aerial photography) (Jensen, 1996).

The following discussion begins with a brief overview of the importance of high spatial resolution aerial photography and satellite imagery to obtain detailed Level III and IV land-cover and land-use information for urban applications and progresses to lower spatial resolution imagery for regional and global applications (mainly associated with Levels I and II). Several sources of remotely sensed data may be available for a given spatial resolution. This discussion deals primarily with publicly available and commercial sources from the United States.

Urban and Suburban Land Cover and Land Use

Many Agenda 21 issues concentrate on urban and suburban areas (Table 2-5). The detailed land-use and land-cover information needed in these settings is derived from high spatial resolution aerial photography or satellite imagery (Table 6-1, Figure 6-2). Since 1994, such companies as Space Imaging and DigitalGlobe have marketed high spatial resolution satellite data (approximately 1 × 1 m to 4 × 4 m) (Annex Box 6-1). Examples of Space Imaging's IKONOS imagery are shown in Figure 6-3.

Ways need to be found to make high spatial resolution imagery accessible to users in Africa. Currently, these data are expensive (Table 6-2), and more affordable, lower spatial resolution imagery is an inadequate substitute in urban environments.

Regional and Global Land Cover

The land cover of much of Africa can be inventoried using medium to coarse spatial resolution satellite imagery (e.g., 20 to 1000 m). Normally this imagery must be multispectral. This section discusses five sources of these data, all of which can be obtained inexpensively. Additional resources can be found at the World Data Center for Remotely Sensed Land Data.[1]

Land-Cover Data Source A: Advanced Very High Resolution Radiometer (AVHRR) Imagery

NOAA's AVHRR is a widely used source of satellite data for natural resource management and early warning systems in Africa. This class of sensor flies onboard NOAA's operational satellites (Annex Box 6-2), and will likely continue operating until 2018 (Annex Box 6-3). AVHRR is a sustained source of low-cost data with a spatial resolution of ~1 × 1 km.

The Global Land Cover Dataset

AVHRR images from 1992 and 1993 are the source for the Global Land Cover dataset. The dataset was compiled for broad use in environmental research and modeling (Loveland et al., 2000). It was developed by IGBP Data and Information Systems Focus 1 activity (Townshend and Skole, 1995)[2] and implemented by the USGS EROS Data Center, the European Commission's Joint Research Centre,

---

[1] <http://edc.usgs.gov/doc/edchome/world/wdcguide.html>.

[2] Funding for the project is provided by the NASA, NOAA, the U.N. Environment Programme, U.S. Environmental Protection Agency, U.S. Forest Service, USGS, and European Space Agency.

TABLE 6-1 Urban and Suburban Applications and the Minimum Remote-Sensing Resolutions Required to Obtain Such Information

	Minimum Resolution Requirements		
Attributes	Temporal	Spatial	Spectral[a]
Land Use/Land Cover			
L1—USGS Level I	5-10 years	20-100 m	V-NIR-MIR-Radar
L2—USGS Level II	5-10 years	5-20 m	V-NIR-MIR-Radar
L3—USGS Level III	3-5 years	1-5 m	Pan-V-NIR-MIR
L4—USGS Level IV	1-3 years	0.25-1 m	Panchromatic
Building and Property Infrastructure			
B1—Building perimeter, area, height, and cadastral information (property lines)	1-5 years	0.25-0.5 m	Pan-Visible
Transportation Infrastructure			
T1—General road centerline	1-5 years	1-30 m	Pan-V-NIR
T2—Precise road width	1-2 years	0.25-0.5 m	Pan-Visible
T3—Traffic count studies (e.g., cars, airplanes)	5-10 min	0.25-0.5 m	Pan-Visible
T4—Parking studies	10-60 min	0.25-0.5 m	Pan-Visible
Utility Infrastructure			
U1—General utility line mapping and routing	1-5 years	1-30 m	Pan-V-NIR
U2—Precise utility line width, right-of-way	1-2 years	0.25-0.6 m	Pan-Visible
U3—Location of poles, manholes, substations	1-2 years	0.25-0.6 m	Panchromatic
Digital Elevation Model (DEM) Creation			
D1—Large scale DEM	5-10 years	0.25-0.5 m	Pan-Visible
D2—Large scale slope map	5-10 years	0.25-0.5 m	Pan-Visible
Socioeconomic Characteristics			
S1—Local population estimation	5-7 years	0.25-5 m	Pan-V-NIR
S2—Regional and national population estimation	5-15 years	5-20 m	Pan-V-NIR
S3—Quality of life indicators	5-10 years	0.25-30 m	Pan-V-NIR
Energy Demand and Conservation			
E1—Energy demand and production potential	1-5 years	0.25-1 m	Pan-V-NIR
E2—Building insulation surveys	1-5 years	1-5 m	TIR
Critical Environmental Area Assessment			
C1—Stable sensitive environments	1-2 years	1-10 m	V-NIR-MIR
C2—Dynamic sensitive environments	1-6 months	0.25-2 m	V-NIR-MIR-TIR
Disaster Emergency Response			
DE1—Pre-emergency imagery	1-5 years	1-5 m	Pan-V-NIR
DE2—Post-emergency imagery	12 hr-2 days	0.25-2 m	Pan-V-NIR-Radar
DE3—Damaged housing stock	1-2 days	0.25-1 m	Pan-V-NIR
DE4—Damaged transportation	1-2 days	0.25-1 m	Pan-V-NIR
DE5—Damaged utilities, services	1-2 days	0.25-1 m	Pan-V-NIR
Meteorological Data			
M1—Weather prediction	3-25 min	1-8 km	V-NIR-TIR
M2—Current temperature	3-25 min	1-8 km	TIR
M3—Clear air and precipitation mode	6-10 min	1 km	WSR-88D Radar
M4—Severe weather mode	5 min	1 km	WSR-88D Radar
M5—Monitoring urban heat island effect	12-24 hr	5-30 m	TIR

[a]Spectral resolution is the extent to which an application requires detection of light within narrow bands of the electromagnetic spectrum such as visible blue, green, and red light (V), a single broad band of visible light (e.g., encompassing both green and red light; Pan), near-infrared (NIR) energy, middle-infrared (MIR), and thermal-infrared (TIR). SOURCE: Jensen and Cowen, 1999.

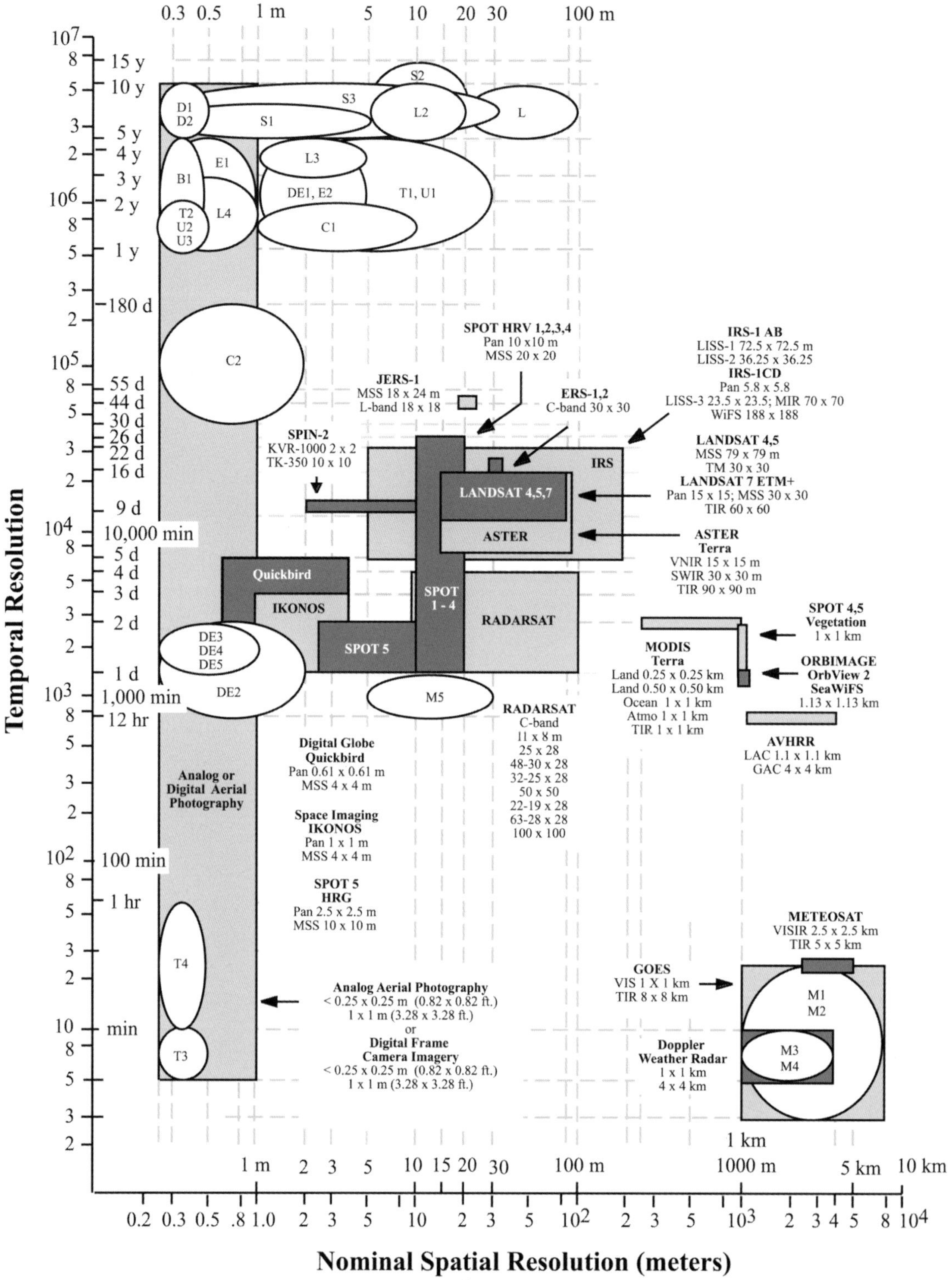

FIGURE 6-2 The relationship between the spatial and temporal resolution of urban and suburban attributes and the spatial and temporal resolution of various aerial and sub-orbital remote-sensing systems. The clear polygons represent the spatial and temporal requirements for selected urban attributes listed in Table 6-1. Gray boxes depict the spatial and temporal characteristics of selected major remote-sensing systems that may be used to extract the required urban information (updated from Jensen and Cowen, 1999).

A

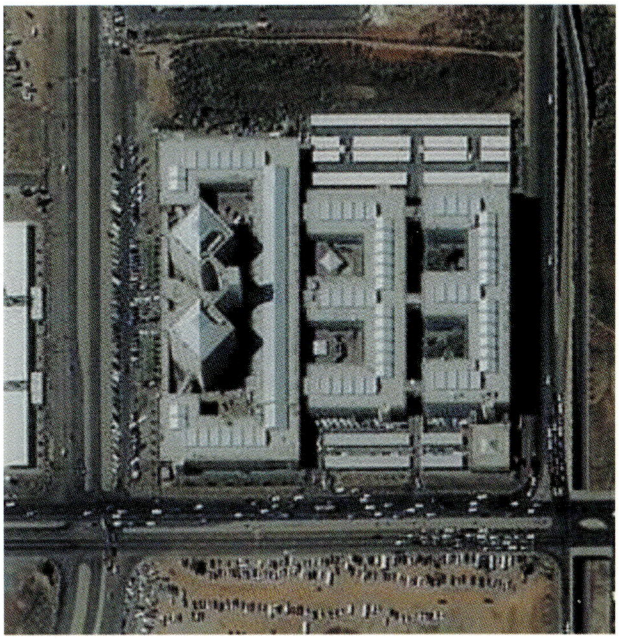

B

FIGURE 6-3 IKONOS 1 × 1 m panchromatic imagery of (a) the Grand Pyramid in Giza obtained on November 17, 1999, and (b) 1 × 1 m pan-sharpened image of a mosque in Abuja, Nigeria, obtained on November 7, 2001 (courtesy of Space Imaging, Inc.).

and the NASA Earth Observing System (EOS) Pathfinder program.

The Global Land Cover dataset is available by continent, including Africa (Figures 6-4 and 6-5). Additionally, there are seven global datasets, each using a different landscape classification:

- Global Ecosystems (Olson, 1994a,b);
- IGBP Land Cover Classification (Belward, 1996);
- U.S. Geological Survey Land Use/Land Cover System (Anderson et al., 1976);
- Simple Biosphere Model (Sellers et al., 1996);
- Simple Biosphere 2 Model (Sellers et al., 1996);
- Biosphere Atmosphere Transfer Scheme (Dickinson et al., 1986); and
- Vegetation Lifeform (Running et al., 1995).

The first version of the dataset was released in 1997[3] and was subjected to a formal accuracy assessment.[4] A revised version is now available, although the accuracy of this version has yet to be formally assessed. Unless protected by copyrights or trade secret agreements, all data generated for the Global Land Cover dataset (source, interpretations, attributes, and derived data) are distributed at cost of filing a user request through the USGS EROS Data Center Distributed Active Archive Center for land processes data (USGS, 2002a).

Tropical Forest Extent

AVHRR data were used by the Tropical Ecosystem Environment Observations by Satellite (TREES) project, a European Commission initiative, to map tropical forest extent.[5] TREES activities were coordinated with those of the Global Land Cover project through the IGBP Data and Information Systems program.

The first phase of TREES produced a baseline assessment of humid tropical forest cover for 1992. Three regional vegetation maps (each at a scale of 1:5,000,000) have been published or are under development: (1) Central Africa (Mayaux et al., 1997), (2) South America (Eva et al., 1998), and (3) continental Southeast Asia (in preparation) (TREES, 2002). The second phase of the project assessed forested area change in the humid tropics. The resulting "Hot Spot Report" (Achard et al., 1998) highlights areas with rapid forest-cover changes. Hot spot maps are available for central Africa (Figure 6-6), West Africa, and Madagascar (TREES, 2002).

---

[3]As an International Geosphere Biosphere Program, Data and Information System, initiative led by the Land Cover Working Group.

[4]This included validation of the land-cover maps by organizations including the Miombo Network (Chapter 7). The product is known to contain some inaccuracies, particularly for cropland which is difficult to map in Africa. Loveland et al. (2000) discuss the accuracy of the dataset.

[5]TREES, part of a project called World Forest Watch involving space agencies worldwide, was initiated during the International Space Year in 1992.

TABLE 6-2 Costs of Remotely Sensed Satellite Imagery

Satellite	Scene Width (km)	Cost per Scene (U.S. $)[a]	Spatial Resolution (m)[b]	Revisit Frequency (days)	Advantages	Disadvantages
IKONOS	11	3,500 Pan 3,500 MSS	1 Pan 4 MSS	1-4	Very detailed imagery; in-orbit programming possible.	Expensive (per km^2); copyright restrictions for sharing data; sensitive to cloud cover.
KVR-1000	40	3,500	1 Pan	Irregular	Detailed imagery; historic data available.	Expensive (per km^2); original data not in digital form; relatively long delivery time; sensitive to cloud cover.
IRS-1C/D	71 140	2,500 Pan 2,500 MSS +SWIR	6 Pan 25 MSS 70 SWIR	24 (12 for C/D couple)	Proven: relatively detailed imagery.	Expensive (per km^2); copyright restrictions; sensitive to cloud cover.
SPOT HRV	60	2,500 Pan 2,000 MSS +SWIR	10 Pan 20 MSS 20 SWIR	26, but shorter frequency possible	Proven; multiple applications; programmable; historic data record available.	Expensive (per km^2); copyright restrictions; sensitive to cloud cover.
Landsat 4-5 TM	185	2,500 for all channels	30 MSS 120 IR	16	Proven; multiple applications; historic data record available; compatible with previous Landsat data for change detection.	Expensive (per km^2); copyright restrictions; long revisit interval; sensitive to cloud cover.
Landsat 7 ETM+	185	600 for all channels	15 Pan 30 MSS 60 IR	16	Proven; multiple applications; compatible with previous Landsat data for change detection; inexpensive (per km^2); no copyright restrictions for sharing data.	Long revisit interval; sensitive to cloud cover.
SPOT Vegetation	2,250	170 for all channels	1,160 MSS	1	Global daily coverage; provides aggregated 10-day average of global vegetation cover.	Only available since 1998; copyright restrictions for sharing data; sensitive to cloud cover.
NOAA AVHRR	2,400	Not applicable	1,100 MSS	0.5 (two satellites)	Data available since 1978; near real-time delivery; NDVI vegetation index data available at low or no cost.	Coarse resolution; sensitive to cloud cover.
DMSP OLS	3,000	Not applicable	550 VIS 2,700 IR	1	Data available since 1978 at low or no cost.	Coarse resolution.

[a]The cost per scene is computed using a rectangle based on the scene width listed in column 2 (unless otherwise indicated).
[b]IR = Infrared, MS = Multispectral, Pan = Panchromatic, SWIR = Short Wave Infrared, VIS = Visible

SOURCE: Adapted from U.S. Institute for Peace (2002)

Land-Cover Data Source B: The Moderate Resolution Imaging Spectroradiometer (MODIS) Sensor

The MODIS sensor onboard NASA's Terra satellite (Annex Box 6-4) measures a wide array of parameters, including land cover.[6] The aim of the Terra research mission is to monitor and document global climate change, land use, land cover, and other factors affecting human habitability (Figure 6-7). Launched in December 1999, the Terra satellite is one of NASA's Earth Observing System satellites. The system-

[6]A comparison of Annex Boxes 6-2 and 6-4 reveals the higher spatial resolution of MODIS over AVHRR in a number of wavelength ranges. MODIS also has a wider array of potential applications to Earth resource issues (see Annex Box 6-4).

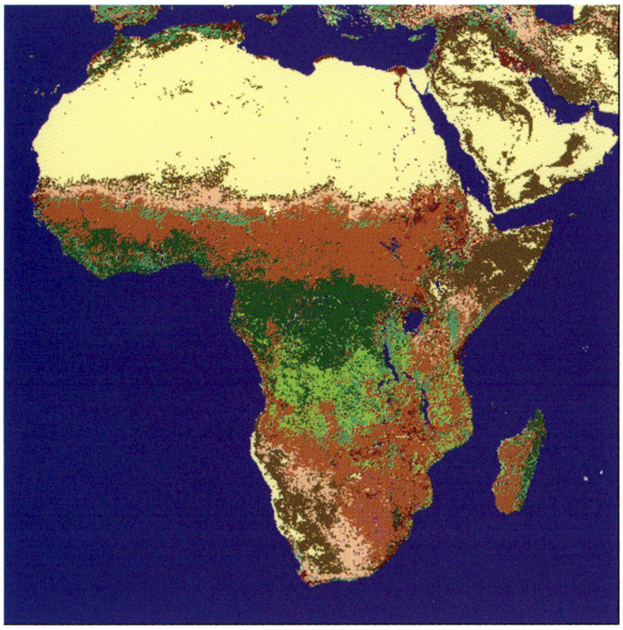

FIGURE 6-4 A much reduced map of the land cover of Africa derived from NOAA AVHRR 1 × 1 km data in 1992-1993 (courtesy U.S. Geological Survey EROS Data Center). The distance (parallel to the Equator) from the tip of Horn of Africa to the west coast off Guinea-Bissau is 7,350 km. Land cover types: pale yellow = barren; light brown = savanna; olive = shrubland; pink = grassland; bright green = deciduous forest; dark green = evergreen forest.

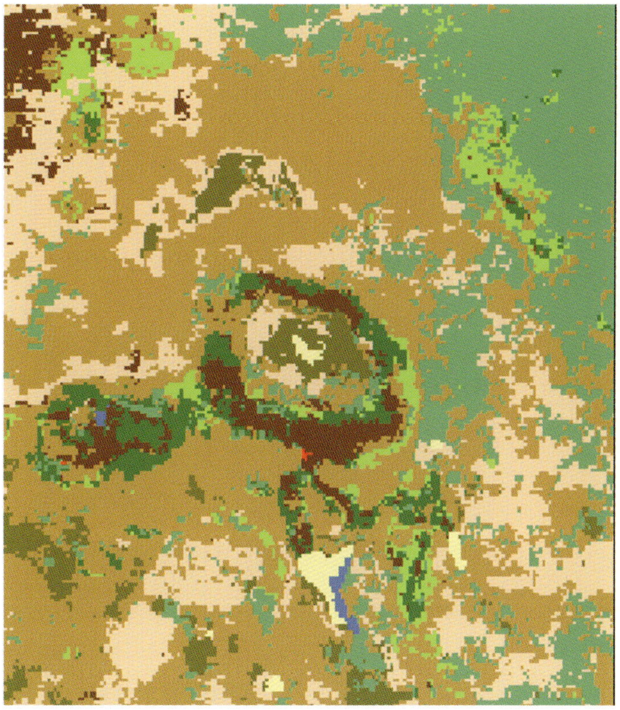

FIGURE 6-5 Full resolution land-cover map of the area centered on Mount Kilimanjaro derived from NOAA AVHRR 1 × 1 km data in 1992-1993 (courtesy U.S. Geological Survey EROS Data Center). The width of the depicted area is 190 km. Land-cover types: pale yellow = barren; light brown = savanna; olive = shrubland; pink = grassland; bright green = deciduous forest; dark green = evergreen forest; red = developed; dark brown = cropland or pasture; pale green = cropland or woodland; blue = water.

atic observations begun with Terra and maintained on other satellites are planned to continue for at least 15 years.

A major mission of the MODIS sensor is to characterize land cover and global primary productivity (Justice et al., 1998). A year (a complete seasonal cycle) is needed to acquire the raw data for each land-cover dataset. The MODIS land-cover product will identify 17 classes of land cover in the IGBP global vegetation classification scheme[7] with a spatial resolution of 1 × 1 km. A global vegetation cover change product is also being developed at a spatial resolution of 250 × 250 m. MODIS data products are being released sequentially, with products for Africa available shortly.

NASA's MODIS science team is engaging African scientists through networks including the Miombo Network (Chapter 7). Additionally, NASA science campaigns such as Safari2000 are working to put MODIS data in the hands of scientists.

Land-Cover Data Source C: Landsat Data

Landsat data have spatial resolutions ranging from 15 × 15 m (Enhanced Thematic Mapper Plus–ETM+) to 79 × 79 m (Multi-Spectral Scanner) (Chapter 5; Annex Box 5-3). As such, Landsat data contain much more spatial information than either AVHRR or MODIS data. The visible, near-infrared, and middle-infrared Landsat Thematic Mapper bands are particularly useful for many vegetation-mapping applications. Landsat data, however, are costly and therefore inaccessible to many potential users unless the data have been purchased, and appropriate sharing arrangements negotiated by a government agency (e.g., NASA [Chapter 5] and NIMA [see below]) or other organizations (e.g., U.N. Food and Agriculture Organization [FAO] [see below]). The tradeoffs between AVHRR or MODIS and Landsat data are primarily between spatial resolution and cost of repeated data collection for change detection.

EarthSat GeoCover Land-Cover Data

In 1999 Earth Satellite Corporation began preparing a Landsat-based land-cover database called "GeoCover-Land

[7]This scheme includes 11 natural vegetation classes, 3 developed land classes, 1 of which is a mosaic with natural vegetation, permanent snow or ice, barren or sparsely vegetated, and water.

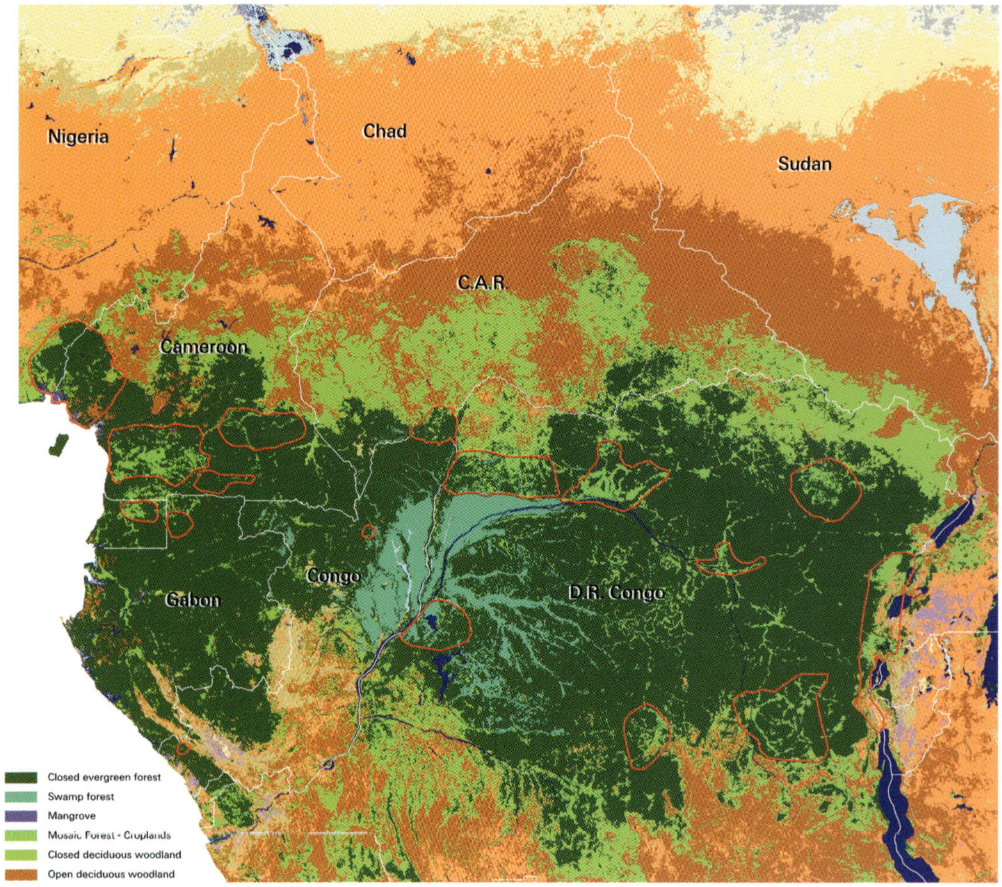

FIGURE 6-6 Forest cover for central Africa and deforestation hot spots (areas circled in red) between 1992 and 1994 derived from NOAA AVHRR data (courtesy of the Tropical Ecosystem Environment Observation by Satellite group). The width of the depicted area is 2,500 km.

Cover" for NIMA (Earth Satellite, 2002). The database for Africa has been completed.[8] For example, a land-cover map of the area centered on Mount Kilimanjaro in East Africa is shown in Figure 6-8. This product was produced from EarthSat's GeoCover-Ortho product (Figure 5-2). The database has a spatial resolution of 30 × 30 m, and contains 13 land-cover classes of Earth's land areas. EarthSat is also processing global frames of Landsat Multi-Spectral Scanner data obtained during the 1970s and global frames of Landsat TM data obtained in the late 1990s and early 2000s.

Africover Land-Cover Mapping Project

Initiated in 1996, FAO's Africover project responded to national requests for assistance in obtaining reliable geographically referenced information on natural resources at national and regional scales. The principal sources of data for the project are Landsat 5 Thematic Mapper and Landsat Multi-Spectral Scanner satellite images. The project aims to create two databases for Africa: one a digital land-cover database, the other a geographic database (including roads and hydrography), at a scale of 1:200,000 (1:100,000 for small countries and specific areas) (FAO, 2002a). The land-cover database can be used for forest and rangeland monitoring, watershed management, biodiversity or climate change studies, and in famine early warning systems.

The first operational module of the Africover initiative covers eastern Africa (including Burundi, Democratic Republic of Congo, Egypt, Eritrea, Kenya, Rwanda, Somalia, Sudan, Tanzania, Uganda). Representing roughly one third of Africa (by area),[9] this module began in 1997 with funding from the Italian and U.S. governments.

The Africover initiative differs from other examples of land-cover monitoring described in this report because it involves Africans in many aspects of data processing. The resultant network of scientists and technicians is linking with related initiatives such as the UN program on Global Terres-

[8]The data, from as close as possible to peak growing season between 1987 and 1993, are available from Earth Satellite Corporation in raster or vector format with a 1.4 ha minimum mapping unit, or in Landsat TM raster format at 0.08 ha.

[9]And covered by 400 Landsat images.

FIGURE 6-7 A land-cover image of the middle and lower Nile from Terra MODIS imagery (courtesy NASA).

trial Observing Strategy (Box 6-1). Land-cover data are derived primarily from on-screen image interpretation in host countries. The land-cover classification is performed manually (as opposed to automatically by a computer) using the "Africover Interpretation and Mapping System," and the "FAO Land Cover Classification System" (Degregorio and Jansen, 2000). In addition to Landsat data, aerial photography and other geographically referenced data are used during the classification process, and subsequently the accuracy of classifications is verified in the field. The land-cover database is accessible through the Africover Database Gateway (FAO, 2002b).

Land-Cover Data Source D: Declassified Remote Sensor Data

Executive Order Number 12951, issued by President William Clinton on February 22, 1995, directed that

> [i]magery acquired by the space-based national intelligence reconnaissance systems known as the Corona, Argon, and Lanyard Missions shall, within 18 months of the date of this order be declassified.

These declassified photographs from U.S. spy satellites are a rich source of historical land-cover and land-use information for many areas in the world (USGS, 1998; Clarke, 1999), including Africa. The satellite photographs may be browsed at no cost using the USGS Global Land Information System and purchased for U.S.$16 to $75 (depending on the size of the photograph) from the USGS EROS Data Center (USGS, 2002b).[10] The photographs date from the late 1950s to early 1970s (Peebles, 1997) and often are the earliest satellite photographic record of an area. They are baseline data with which to compare later images for change detection. In an application of some of the hundreds of photographs collected over Africa, Tappan et al. (2000) used Argon and Corona data from 1963 to map historical agricultural practices in Senegal.

Corona, Argon, and Lanyard Images

In 1959 the United States launched Corona, its first reconnaissance satellite (Day et al., 1998). In it's ninth and first successful mission Corona provided more photographic coverage of the Soviet Union than all previous U-2 spy plane missions combined. Between 1960 and 1972 the spatial resolution of a sequence of Corona satellites improved from 25-40 ft to 4.5-6 ft (Ruffner, 1995; McDonald, 1997) (Table 6-3).[11] By 1972 Corona missions, which were followed by the Argon and Lanyard missions, acquired over 800,000 images of Earth (Clarke, 1999) (e.g., Figure 6-9).

Land Cover Data Source E: Space Photography

Since 1961, NASA astronauts have used hand-held cameras to capture approximately 340,000 photographs of Earth (Lulla et al., 1994). Many of these photographs have spatial resolutions similar to Landsat Thematic Mapper and Terra MODIS data.[12] They indicate land cover and, in areas with repeat coverage, any change in this parameter since 1961.

Space photography was formalized in the Space Shuttle Earth Observation Photography program and is continued on the International Space Station using digital imaging systems in addition to cameras. Images with a spatial resolution

---

[10]The U.S. National Archives and Records Administration holds the original negatives as well as technical mission-related documents that include the orbit parameters for each mission.

[11]The code word "keyhole," abbreviated to "KH," referred to the camera systems on these reconnaissance programs. KH-1, KH-2, KH-3, and KH-4 were Corona sensors; Argon's camera was KH-5, and Lanyard's camera was KH-6. All KH-4 satellites (1962–1972) contained twin panoramic cameras that could obtain stereoscopic photographs, useful for viewing the land surface in three dimensions.

[12]Metadata for the space photography consists of latitude and longitude of the center of the photograph, position of spacecraft, degree of cloud cover, description of observable features within the picture, and geographical information about the political location. Coverage is primarily between 28 degrees N and S latitudes but up to 57 degrees N and S latitudes. Stereoscopic coverage is available for a number of areas, enabling construction of digital elevation models in certain instances.

BOX 6-1
The Global Observations of Forest-Global Observations of Land Cover Dynamics Program

The Global Observations of Forest–Global Observations of Land Cover Dynamics (GOFC-GOLD) program is an international activity providing space-based and in situ observations of forest and other vegetation cover for (1) sustainable management of terrestrial resources and (2) obtaining an accurate understanding of the terrestrial carbon budget. It operates under the auspices of the U.N. program on Global Terrestrial Observing Strategy.

The GOFC-GOLD program works to accomplish its objectives by (1) providing a forum for users of satellite data to discuss their needs and for producers to respond through improvements to their programs; (2) providing regional and global datasets containing information on location of different forest types, major changes in forest cover, and the biological functioning of forests (to help quantify the contribution forests make as absorbers and emitters of greenhouse gases); (3) promoting international networks for data access, data sharing, and collaboration; and (4) stimulating the production of improved datasets.

The program is partnered with the FAO and its Africover project, and operates in Africa through two major networks of local participants. One network coordinates scientists from central Africa in the application of remote-sensing data for forest-cover change analysis and methods for measuring and inventorying forest resources. In southern Africa the program is implemented through collaborative links with the Miombo Network (Chapter 7), a network of scientists in the region focused on fire detection and land-use and ad cover change.

The GOFC-GOLD implementation strategy is to demonstrate operational forest monitoring at regional and global scales by conducting pilot projects and developing prototype products within three themes: (1) forest cover characteristics and changes, (2) forest fire monitoring and mapping, and (3) forest biophysical processes.

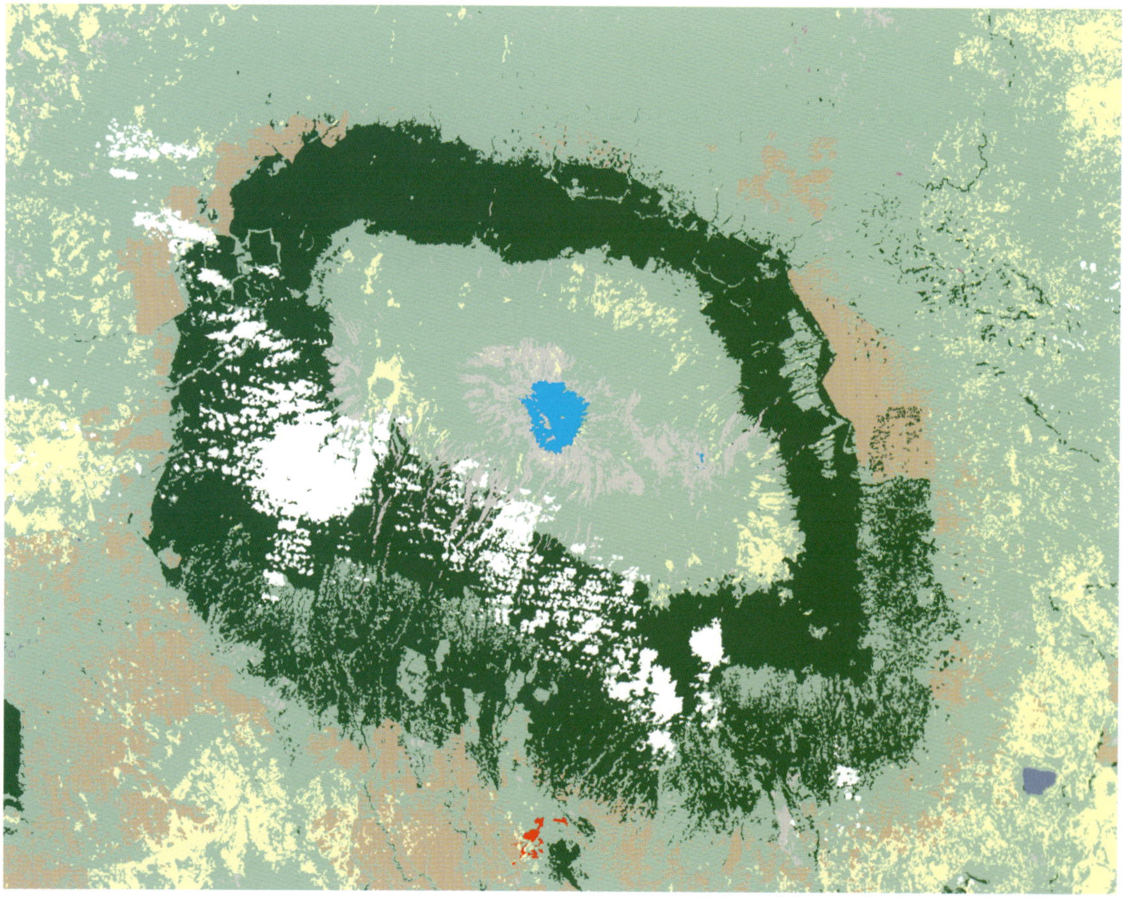

FIGURE 6-8 A land-cover map of Mount Kilimanjaro from the EarthSat GeoCover land-cover mapping project. The map contains 13 classes and was derived from Landsat Thematic Mapper imagery obtained in the early 1990s. Compare this map with the original imagery found in Figure 5-2 (courtesy of Earth Satellite Corporation). The width of the depicted area is 110 km. Dark greens represent the most dense vegetation, pale greens are sparse vegetation, yellows are no vegetation, and browns are croplands.

TABLE 6-3 Summary of Declassified Satellite Missions

Orbital Satellite System	Dates of Operation	Ground Resolution	Frames
KH-1 (Single camera)	June 1959-September 1960	25-40 ft.	1,432
KH-2 (Single camera)	October 1960-October 1961	25-40 ft.	7,246
KH-3 (Single camera)	August 1961-January 1962	25-40 ft.	9,918
KH-4 (2 cameras)	February 1962-December 1963	25-40 ft	101,743
KH-4A (2 cameras)	August 1963-August 1969	9 ft	517,688
KH-4B (2 cameras)	September 1967-May 1972	4.5-6 ft	188,526
KH-5 (Global coverage mapping camera)	February 1961-August 1964	460 ft	38,578
KH-6 (Panoramic camera)	July 1963	6 ft	<910

SOURCE: Ruffner (1995); McDonald (1997); USGS (1998).

of 6 × 6 m are being obtained from the International Space Station (Robinson and Evans, 2002). Approximately 14 percent of NASA's space shuttle photographs cover parts of Africa (Figures 6-10 and 6-11). The photographs are routinely digitized and are in the public domain. NASA's Johnson Space Center maintains all cataloged space shuttle Earth photography, and digital files (in compressed format) of all the photographs may be accessed through the following web sites: <http://earth.jsc.nasa.gov> or <http://eol.jsc.nasa.gov>. Slides, prints, or high-resolution uncompressed digital files are available for the cost of processing.

FIGURE 6-9 A much reduced photomosaic of Africa produced from Argon (KH-5) photography obtained from 1961 to 1964 (courtesy U.S. Geological Survey and Keith Clarke, Project Corona at the University of California at Santa Barbara). The distance (parallel to the Equator) from the tip of the Horn of Africa to the west coast off Guinea-Bissau is 7,350 km.

The Future of Land-Cover Data Sources for Africa

There are low-cost sources of coarse and medium spatial resolution land-cover information for Africa. These come from sensors that include AVHRR (1 × 1 km), MODIS (1 × 1 km to 250 × 250 m), and Landsat satellite sensors (79 × 79 m to 15 × 15 m). The resultant datasets include Global Land Cover (AVHRR), TREES (AVHRR), GeoCover Land Cover (Landsat), and Africover (Landsat). Such datasets are valuable resources for natural resource management and development planning in rural areas. Similar datasets can be constructed in the future for change detection as long as there is continued flow of data from AVHRR, MODIS, and Landsat (or their equivalents).

Without some way to assure data continuity (NRC, 1995), investments by development organizations in training and capacity building will be less useful than they could be. And without assurances that these investments will be useful in the future, it will be more difficult for African governments to invest in their own capacity and infrastructure. Changes in data access policy, data cost, or the elimination of an observation program create uncertainties about long-term benefits of international programs to Africans.

There are two areas in particular in which U.S. government agencies should contribute to data continuity.

1. **Until at least 2018 NASA, NOAA, and DOD should carry out their plan for the National Polar-orbiting Operational Environmental Satellite System to ensure that it supplies relatively coarse spatial and high temporal frequency observations (such as the AVHRR follow-on) that are necessary for a multitude of applications in Africa and elsewhere.**
2. **NASA and USGS should take measures to ensure that the Landsat data continuity mission(s) provides long-term continuous data, perhaps through making the Landsat program an operational system for land observations, to support sustainable**

development and natural resource management in Africa and elsewhere. NASA should also ensure that sensors on its Terra and Aqua satellites (e.g., MODIS, ASTER, AMSR-E) continue to provide data for meteorological and land observation applications.

Declassified spy satellite images (e.g., Corona, Argon, Lanyard) and space photography are sources of early (1960s onward) land-cover information against which later sources can be compared for change detection. What is lacking is a low-cost source of very high spatial resolution, up-to-date imagery for urban areas in Africa. A number of Agenda 21 issues center on urban areas, and "image grants" would help to inventory and map the continually changing characteristics of urban infrastructures. Therefore, **USAID should consider purchasing very high spatial resolution images (i.e., < 1 × 1 m) on a regular basis (at 5 to 10 year intervals) for urban areas in Africa and donating them to African organizations to ensure continuity of the data source and change detection.** The imagery might include airborne analog or digital photography or satellite-derived high-resolution imagery. The areas surveyed could be requested by African organizations on the basis of importance of problem and technical and organizational capacity to use the data. One model for this concept is the U.S. Science Data Buy (Box 5-3).

THE CONDITION OF VEGETATION AND HYDROLOGIC RESOURCES

Measurements of the condition of vegetation, soil, and water resources (biophysical measurements) are needed to

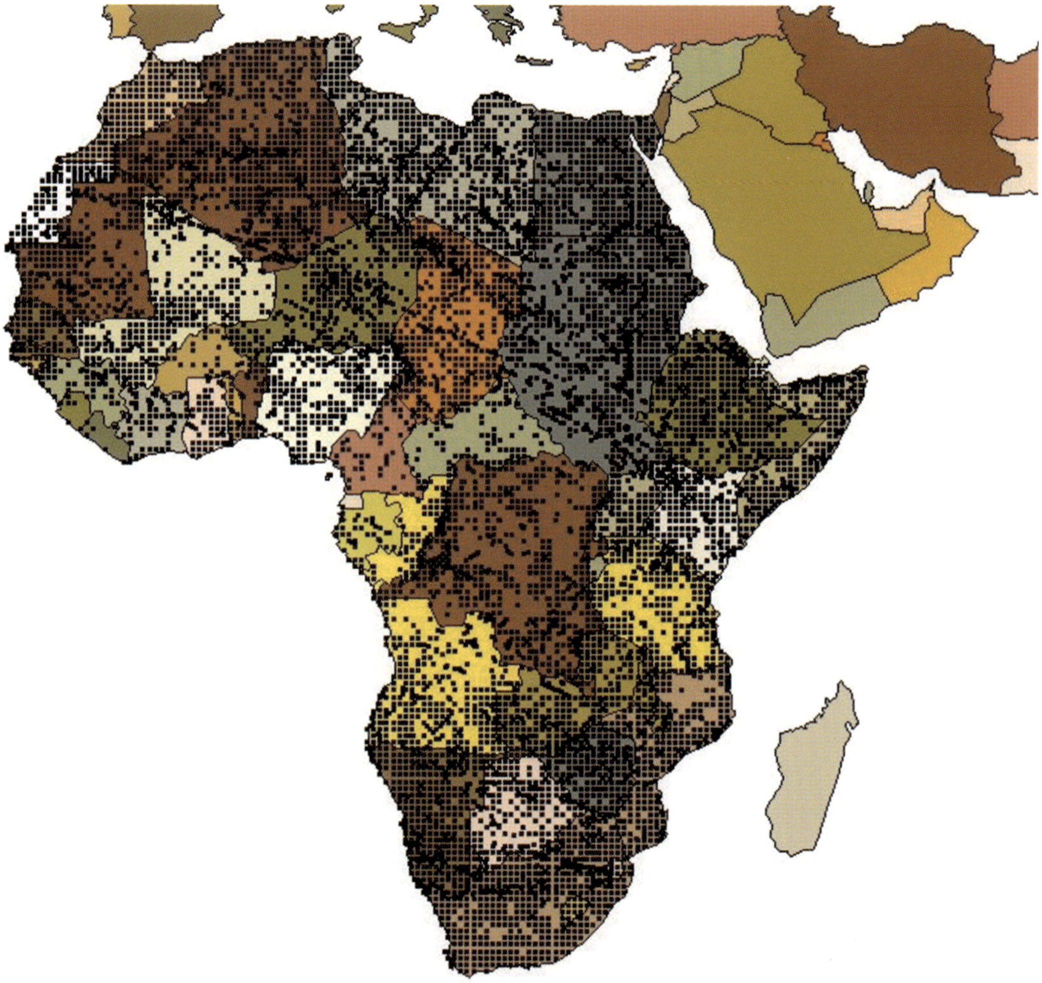

FIGURE 6-10 The geographic distribution of 54,866 space shuttle photographs obtained over Africa prior to 2002 (courtesy Kamlesh Lulla, NASA Johnson Space Center). The distance (parallel to the Equator) from the tip of Horn of Africa to the west coast off Guinea-Bissau is 7,350 km.

FIGURE 6-11 Natural color space shuttle photography of Mount Kilimanjaro in Tanzania (courtesy Kamlesh Lulla, NASA Johnson Space Center). The width of the depicted area is 280 km.

address many environmental and human-induced challenges. For example, these data are pertinent to Agenda 21 issues, including integrated planning and management of land resources, combating deforestation, combating desertification and drought, and promoting sustainable agriculture and rural development (Table 2-5). This section examines the capability of remote-sensing technology to measure (1) vegetation condition, including fire, and (2) rainfall.

Vegetation Condition

The health and productivity of African ecosystems (e.g., forests, rangeland, and cropland) can be measured using vegetation indexes such as the Normalized Difference Vegetation Index, the Enhanced Vegetation Index, the Leaf Area Index, and Fraction of Photosynthetically Active Radiation. Additionally, the occurrence of vegetation on fire can be detected remotely.

Vegetation Indexes

Normalized Difference Vegetation Index

The Normalized Difference Vegetation Index (NDVI)[13] is used routinely to assess vegetation condition (e.g., Tappan et al., 1992). The NDVI indicates the amount and vigor of vegetation (Prince and Justice, 1991; Tucker et al., 1991; ADDS, 2002), pointing to potential food security problems and drought. NDVI is derived (Annex Box 6-5) using data from NOAA's AVHRR sensor (Annex Boxes 6-2 and 6-3) and the European METEOSAT remote-sensing system.

The Famine Early Warning System Network (FEWS NET) uses NDVI to determine the distribution of vegetation condition across Africa every 10 days (Figure 6-12). NDVI images are used to contrast the current condition of vegetation with that of previous times to detect anomalies that could guide famine relief efforts.

FEWS NET has built a valuable archive of NDVI in Africa over the period from 1982 to present. The archived data are not without flaws, however (Box 6-2). The planned National Polar-orbiting Operational Environmental Satellite System (NPOESS) (Annex Box 6-3) will carry an improved version of the AVHRR sensor that should solve some of the problems working with NOAA AVHRR imagery. FEWS NET NDVI data are distributed to the public through the U.S. Geological Survey's Africa Data Dissemination Service (ADDS, 2002).

Enhanced Vegetation Index

The Enhanced Vegetation Index is derived from atmospherically corrected MODIS hyperspectral data from NASA's Terra satellite (Figure 6-13) (Annex Box 6-4). MODIS data have higher spatial resolution (up to 250 × 250 m) than AVHRR data and greater spectral sensitivity.

Leaf-Area Index and Fraction of Photosynthetically Active Radiation

The interaction of sunlight with a forest canopy or other vegetation influences climate. Knowing how much light is absorbed and distributed among the canopy, understory, and ground reveals the functional health and productivity of forests, rangelands, and croplands, and improves predictions of energy flow between the land surface and the atmosphere.

Leaf Area Index (LAI) and Fraction of Photosynthetically Active Radiation (FPAR) measurements indicate the green leaf area and how much sunlight the leaves are absorbing. Therefore, they aid in understanding interactions of sunlight with vegetation. For example, LAI and FPAR can be used to assess the grazing potential of rangelands. Both indexes can be derived from MODIS satellite imagery (Annex Box 6-4) (Knyazikhin, 2002) (Figure 6-14). Daily measurements are combined at weekly intervals into maps that show leaf area and absorbed sunlight for every square kilometer of Earth's land surface during the period. These data products are available less than one month after collection (NASA, 2002a).

Remote Detection of Fire

Naturally induced (e.g., by lightening) and human-induced fires can be remotely detected. Two sources of data for fire detection are the Defense Meteorological Satellite

[13]Annex Box 6-5 contains more details on the NDVI calculation.

FIGURE 6-12 Ten-day composite Normalized Difference Vegetation Index (NDVI) map of Africa derived from NOAA AVHRR imagery in February 2002 (dekad 2). The image is composed of 8 × 8 km pixels. The darker colors indicate denser vegetation (courtesy U.S. Geological Survey EROS Data Center International Program, FEWS NET, USAID, NOAA, and Global Inventory Monitoring and Modeling Studies at NASA). The distance (parallel to the Equator) from the tip of the Horn of Africa to the west coast off Guinea-Bissau is 7,350 km.

BOX 6-2
Correcting AGRHYMET's NDVI Archive

FEWS NET is one of the most successful demonstrations of the value of remotely sensed imagery in Africa. However, AGRHYMET in Niger found the quality of the NDVI data to be inconsistent, rendering them unreliable for desertification monitoring and drought early warning. Problems included imprecise geographic registration (up to 5 km off) temporal gaps (with only the growing season covered in most years), and anomalously low NDVI values. Consequently, AGRHYMET requested technical assistance from the U.S. government (USGS and NASA) to correct its NDVI archive and improve its operational NDVI capability.

A long-term, reliable NDVI archive will permit AGRHYMET to conduct detailed analyses of desertification, one of its obligations under the U.N. Convention to Combat Desertification, in addition to improving early warning information routinely distributed by AGRHYMET. AGRHYMET staff conduct the computer processing with onsite assistance from USGS and NASA staff. Subsequently, USGS and AGRHYMET are posting the data on their Web sites for public access (<http://edcintl.cr.usgs.gov/adds/index.php> and <http://www.agrhymet.ne>, respectively).

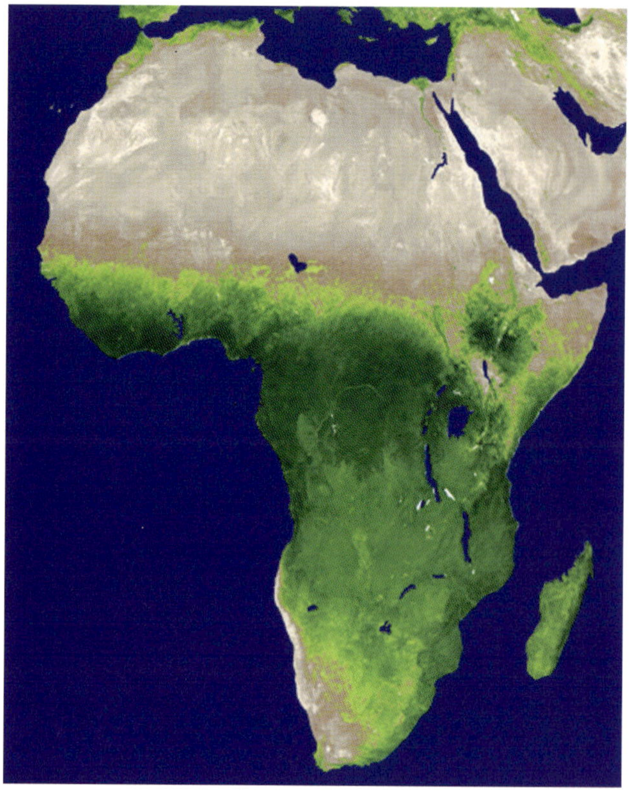

FIGURE 6-13 Enhanced Vegetation Index map of Africa derived from MODIS hyper-spectral imagery collected between February 21 and April 21, 2001 (courtesy of NASA). The darker colors represent denser vegetation. The distance (parallel to the Equator) from the tip of the Horn of Africa to the west coast off Guinea-Bissau is 7,350 km.

Program's (DMSP) Operational Linescan System products (Annex Box 5-6) (Cahoon et al., 1992), and NASA's MODIS sensor (Annex Box 6-4).

The DMSP fire products are generated by overlaying the lights detected from a single orbit on a reference database of stable lights (e.g., city lights) to reveal new sources of light, such as fires (Elvidge et al., 2002) (Figure 5-12). The MODIS products include (1) surface temperature (fires are hot!) and (2) fire occurrence and burn scars (e.g. Figure 6-15) (NASA, 2002b). Fire occurrence data are produced daily (or, for rapid-response products, more frequently).

Rainfall

The livelihoods of the majority of Africans are intimately related to rainfall, which varies greatly from year to year and from place to place. Research over the last three decades has made it possible to obtain rainfall rates using remote passive microwave sensors (Annex Box 6-6). Two sources of rainfall data are the DMSP Special Sensor Microwave/Imager (SSM/I) and NASA and the National Space Development

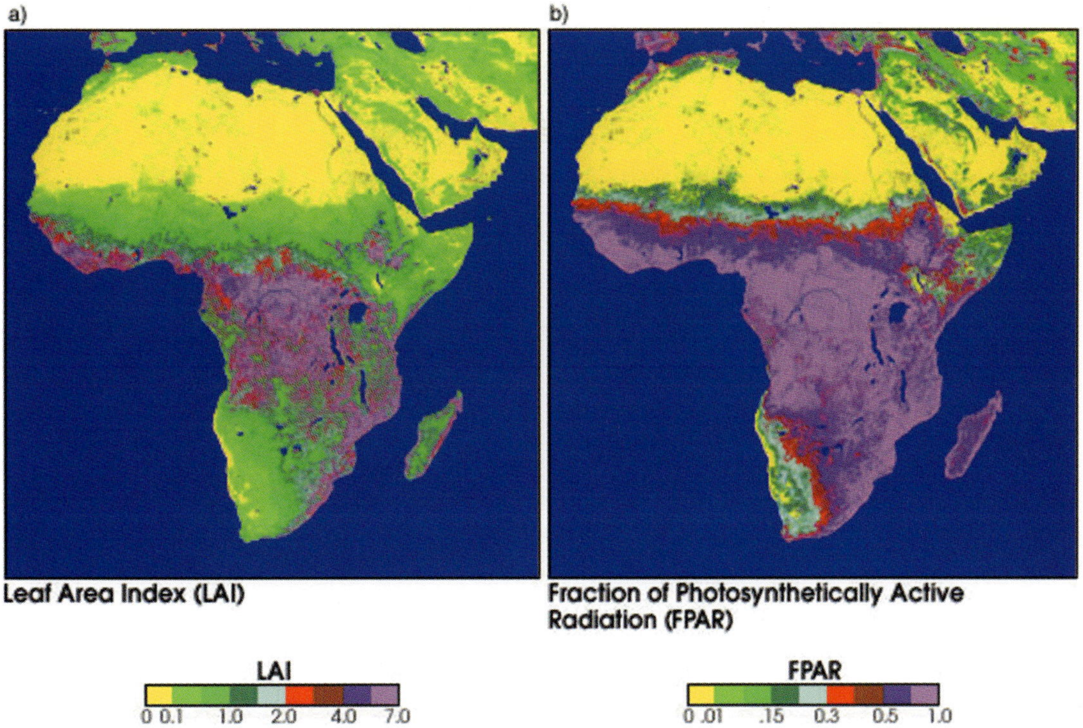

FIGURE 6-14 (a) Leaf Area Index and (b) Fraction of Photosynthetically Active Radiation maps of Africa derived from MODIS hyperspectral imagery (courtesy of NASA). The distance (parallel to the Equator) from the tip of the Horn of Africa to the west coast off Guinea-Bissau is 7,350 km.

Agency of Japan's Tropical Rainfall Measuring Mission (TRMM) (Hou et al., 2001).[14]

Rainfall Detection by the Defense Meteorological Satellite Program

NOAA developed methods for estimating rainfall rates from the SSM/I onboard DMSP satellites through calibration with surface rainfall measurements. Since July 1987, global monthly rainfall data have been produced in 100 × 100 km and 250 × 250 km grids (Ferraro, 1997; Li et al., 1998). These data are available from the National Climatic Data Center Satellite Data Services Division at a cost per orbit of $8 plus cost of media (from <dmsp@ngdc.noaa.gov>).

As a contribution to the USAID-funded FEWS NET, NOAA's Climate Prediction Center also developed a program for rainfall estimation for Africa. The program uses data from numerous sources, including thermal infrared imagery from the European geostationary meteorological satellite (METEOSAT), passive microwave data from the Advanced Microwave Sounding Unit on NOAA's Polar-orbiting Operational Environmental Satellites (POES), DMSP SSM/I data, and rain gauge data retrieved with the World Meteorological Organization Global Telecommunication System. All of this information is processed and made available daily and in 10-day, monthly, and seasonal summaries (Climate Prediction Center, 2002)

Rainfall Detection by the Tropical Rainfall Measuring Mission

The TRMM was launched in 1997 and continues today (Annex Box 6-6). The mission's passive microwave sensor (the TRMM Microwave Imager, TMI) supplies quantitative rainfall information at a spatial resolution of 5 to 45 km. Figure 6-16 depicts a one-month average of rainfall measurements acquired during January of 1998

A new series of experimental near-real-time precipitation estimates is available for latitudes between 50 degrees north and 50 degrees south within about six hours of observation. The data products include a TRMM-calibrated merger of all available TMI and SSM/I precipitation estimates, available in three-hour accumulations (NASA, 2002c).

[14]A third potential source that uses this technology is NASA's Aqua satellite, which was launched on May 4, 2002.

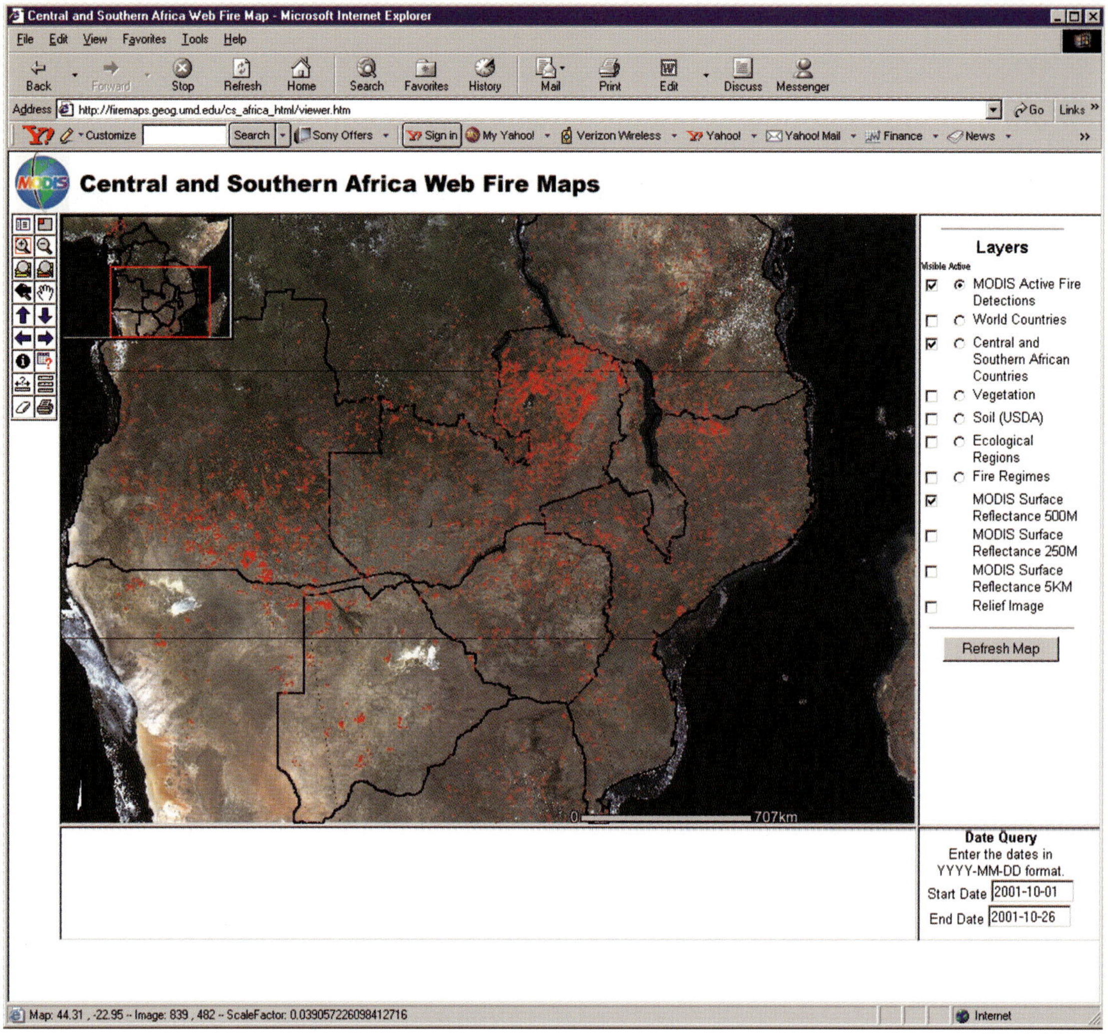

FIGURE 6-15 Results of querying the MODIS Land Science Team Fire Web site (NASA, 2002b). The image shows fires (in red) during October 2001 overlaid on a vegetation surface reflectance image (courtesy NASA).

DATA FOR MANAGING HUMAN HEALTH

Human health is a major challenge for African societies and economies. Disease disrupts families, education, and the workforce. Of the 40 million people worldwide who are infected with AIDS or HIV, about 25 million are in Africa, and the number of Africans infected each year from AIDS-related tuberculosis is about 10 million (WHO, 2001). Additionally, there are approximately 110 million clinical cases of malaria worldwide per year, and over 80 percent of these occur in sub-Saharan Africa.

Although raw numbers on disease incidence are valuable to decision-makers, the distribution and rate of diffusion of disease relates to complex interactions among multiple factors, many of which are geographic (e.g., climate, vegetation, topography, elevation, demography, poverty). A geographic information system facilitates the integration and analysis of these diverse data layers (e.g., Box 6-3) and planning for distribution of medical supplies, assistance, and food. GIS also is a tool for addressing the spread of diseases.

Many of the factors that influence the spread of disease can be mapped using remotely sensed data. NASA's Center for Health Applications of Aerospace Related Technologies (CHAART) generates data that illustrate links between disease and such factors as vegetation that can be remotely sensed (Table 6-4). CHAART evaluates existing and planned remote sensor systems enabling human health scientists to determine relevant data for epidemiological, entomological, and ecological research. It also develops remote-sensing-based models of disease transmission risk (Beck et al., 2000). CHAART conducts several research projects in Africa that apply remotely sensed data (mainly from Landsat Thematic Mapper images) to monitor and predict disease (e.g., Table 6-5).

People are using remotely sensed images and data in GISs for monitoring and evaluating factors associated with disease. They are using satellite instruments for mapping, sur-

BOX 6-3
Controlling Schistosomiasis in Africa

Schistosomiasis is a snail-borne disease. The ability to identify this health threat and monitor the disease enables public health officials to take preventive measures (e.g., vector control). Often the identification of infected human hosts and vector snails depends on labor-intensive ground survey methods for data collection. This method introduces inconsistencies that lead to inaccuracies. By contrast, satellite remote-sensing methods make it possible to obtain standardized data over large geographic areas (Abdel-Rahman et al., 2001). As a result there is increasing interest in these methods for health-related applications.

Remotely sensed data (from NOAA's AVHRR sensor) are being used in spatial decision-support systems to manage control programs for schistosomiasis in Africa. These efforts include a program in the Lake Victoria region building on Malone et al.'s (2001) work in East Africa. The Lake Victoria program also benefits from experiences during a four-year effort in Egypt in which a schistosomiasis risk model was developed for the Ministry of Health (Abdel-Rahman et al., 2001). The model enables the ministry to make more accurate decisions in its program of controlling the spread of schistosomiasis. This GIS-based model, along with the data, constitutes the decision-support system. Two sources of remotely sensed data were used. First, diurnal temperature range and a vegetation index (NDVI) were estimated from NOAA AVHRR data. Second, Landsat Thematic Mapper imagery was used to generate a base map. These data were integrated in a GIS with a database of schistosomiasis prevalence, ground survey results on soil type and salinity, and thematic information from 1:250,000 and 1:10,000 paper maps. From this study it became clear that remote-sensing could extend the capability of the ministry to manage schistosomiasis in Egypt.

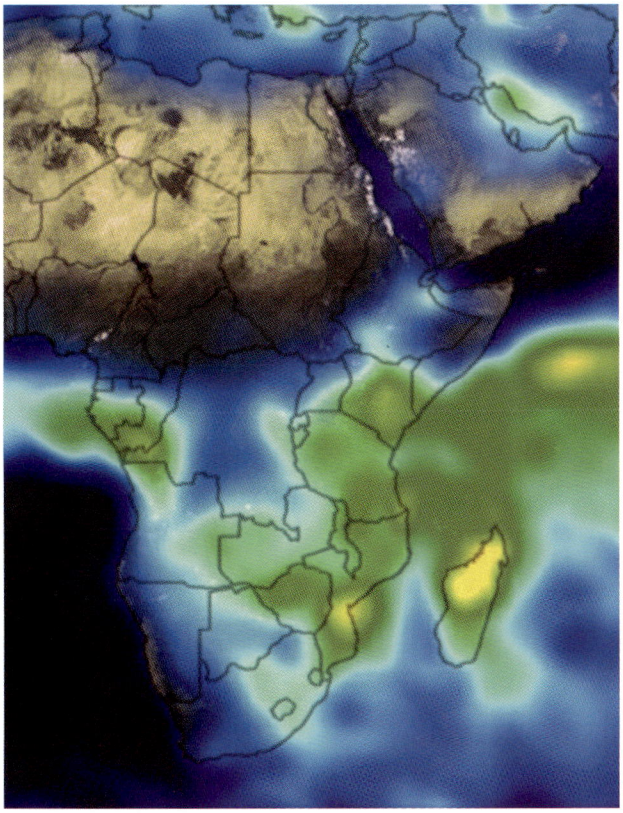

FIGURE 6-16 Average rainfall during January 1998 from the Tropical Rainfall Measuring Mission passive microwave sensor. Low rainfall is indicated by light blue and heavy rainfall by orange and red (courtesy of NASA and the National Space Development Agency of Japan).

veillance, prediction, and control of disease transmission. Moreover, they draw links between environmental variables and disease. As the availability of and access to data and decision-support tools increases, geographic information will become more prominent in efforts to control disease and protect human health in Africa.

COORDINATION AMONG DATA PRODUCERS AND USERS

Moving beyond the current state of the art in the application of geographic data in Africa will require greater attention to coordination among data providers, development assistance agencies, and the science community and end-users in Africa. Already the requirements for the next generation of remote-sensing systems are being defined or developed in many parts of the world, yet there appears to be little dialog between the space agencies and the development assistance agencies, and even less input from potential end users of the data in Africa. Few of the geographic data generation programs now in place have a formal process by which lessons learned in the application of existing data for decision-making are fed back into the definition of future observation and data system requirements, particularly in government science agencies. Consequently, **data providers, U.S. government agencies, and partners should work closely with African organizations to define and integrate the data needs of African users into future data-gathering missions, and to maximize efficiency of new programs through a coordinated approach.** As an added benefit, this dialog will allow users to express their data processing needs.

TABLE 6-4 Links Between Disease and Factors That Can Be Remotely Sensed

Factor	Disease	Mapping Opportunity
Vegetation/crop type	Malaria	Breeding/resting/feeding habitats; crop pesticide vector resistance
	Schistosomiasis	Agricultural association with snails; use of human fertilizers
	Trypanosomiasis	*Glossina* habitat (forests, around villages, depending on species)
	Yellow fever	Reservoir (monkey) habitat
Vegetation green-up	Malaria	Timing of habitat creation
	Rift Valley fever	Rainfall
	Trypanosomiasis	*Glossina* survival
Deforestation	Malaria	Habitat creation (for vectors requiring sunlit pools); habitat destruction (for vectors requiring shaded pools)
	Yellow fever	Migration of infected human workers into forests where vectors exist; migration of disease reservoirs (monkeys) in search of new habitat
Forest patches	Yellow fever	Reservoir (monkey) habitat; migration routes
Flooding	Malaria	Mosquito habitat
	Rift Valley fever	Breeding habitat for mosquito vector
	Schistosomiasis	Habitat creation for snails
Permanent water and wetlands	Filariasis	Breeding habitat for *Mansonia* mosquitoes
	Malaria	Breeding habitat for mosquitoes
	Schistosomiasis	Snail habitat
Canals	Malaria	Dry season mosquito-breeding habitat; ponding; leaking water
	Schistosomiasis	Snail habitat

SOURCE: Adapted from Beck et al. (2000).

TABLE 6-5 Research Using Remotely-Sensed Data to Map Disease Vectors

Disease	Vector	Location	Sensor[a]	Reference
Dracunculiasis	*Cyclops* spp.	Benin	Landsat TM	Clarke et al, 1990
	Cyclops spp.	Nigeria	Landsat TM	Ahearn and De Rooy, 1996
Filariasis	*Culex pipiens*	Egypt	AVHRR	Hassan et al., 1998a
	Culex pipiens	Egypt	Landsat TM	Hassan et al., 1998b; Cross et al., 1996
Malaria	*Anoepheles* spp.	Gambia	AVHRR, Meteosat	Thomson et al., 1997; Beck et al., 1994
		Kenya	RADARSAT-1	Kaya et al., 2002
Rift Valley fever	*Aedes* & *Culex*. spp	Kenya	AVHRR	Linthicum et al., 1990; Pope et al., 1992
	Culex. spp.	Kenya	Landsat TM, Synthetic Aperture Radar	Linthicum et al., 1994
	Culex. spp.	Senegal	SPOT, AVHRR	Malone et al., 1994
Schistosomiasis	*Biomphalaria* spp.	Egypt	AVHRR	Rogers, 1991
Trypanosomiasis	*Glossina* spp	Kenya, Uganda	AVHRR	Kitron et al., 1996
	Glossina spp	Kenya	Landsat TM	Rogers and Randolph, 1991
	Glossina spp	West Africa	AVHRR	Rogers and Williams, 1993
	Glossina spp	Africa	AVHRR	Robinson et al., 1997
	Glossina spp	Southern Africa	AVHRR	CEOS, 1995

[a]TM = Thematic Mapper; AVHRR = (NOAA's) Advanced Very High Resolution Radiometer; SPOT = Système Pour l'Observation de la Terre.

SOURCE: Adapted from Beck et al. (2000)

For example, the option for users in developing countries to obtain geographic data in processed or raw form from government and private data sources will allow flexibility in the required level of geospatial capacity to use the data.

SUMMARY

Many types of thematic geographic data such as land cover, biophysical data, and some data for managing human health are, with the exception of very high spatial resolution urban land-cover data, available at low cost for addressing Agenda 21 issues in Africa. There are many existing applications of these data. Continuity of the data sources or their equivalents, options for raw and processed data, and coordination among data providers and users are crucial for continued and expanded use of geographic data for sustainable development in Africa.

The next chapter explores how people manage, analyze, and subsequently integrate geographic data into the decision-making process.

REFERENCES

Abdel-Rahman, M. S., M. M. El-Bahy, J. B. Malone, R. A. Thompson, and N. M. El Bahy. 2001. Geographic information systems as a tool for control program management for schistosomiasis in Egypt. Acta Tropica 79:49-57.

Achard, F., H. Eva., A. Glinni, P. Mayaux, T. Richards, and H. Stibig (eds). 1998. Identification of Deforestation Hot Spot Areas in the Humid Tropics. TREES Publications Series B. Research Report no. 4. Ispra, Italy: Joint Research Centre, European Commission.

ADDS (Africa Data Dissemination Service). 2002. Africa Data Dissemination Service homepage. Available at <http://edcintl.cr.usgs.gov/adds/NDVIPaper.php>. Accessed June 27, 2002.

Ahearn, S. C., and C. De Rooy. 1996. Monitoring the effects of dracunculiasis remediation on agricultural productivity using satellite data. International Journal of Remote Sensing 17:917-929.

Anderson, J. R., E. E. Hardy, J. T. Roach, and R. E. Witmer. 1976. A land use and land cover classification system for use with remote sensor data. U.S. Geological Survey Professional Paper 964. Washington, DC: U.S. Government Printing Office.

Beck L. R., M. H. Rodríguez, S. W. Dister, A. D. Rodríguez, E. Rejmánková, and A. Ulloa. 1994. Remote sensing as a landscape epidemiologic tool to identify villages at high risk for malaria transmission. American Journal of Tropical Medicine Hyg. 51:271-280.

Beck, L. R., B. M. Lobitz, and B. L. Wood. 2000. Remote Sensing and Human Health: New Sensors and New Opportunities, *Emerging Infectious Diseases*. Available at <http://www.cdc.gov/ncidod/eid/vol6no3/beck.htm>. Accessed June 5, 2002.

Belward, A.S. (ed). 1996. The IGBP-DIS Global 1 km Land Cover Data Set (DISCover): Proposal and Implementation Plans. IGBP-DIS Working Paper No. 13. Toulouse, France: IGBP-DIS Office. Meteo-France.

Cahoon, D. R. Jr., B .J. Stocks, J. S. Levine, W. S. Cofer, III, and K. P. O'Neill. 1992. Seasonal Distribution of African Savanna Fires. Nature 359:812-815.

CEOS (Committee on Earth Observation Satellites). 1995. Coordination for the Next Decade (1995 CEOS Yearbook). European Space Agency. Smith System Engineering Ltd, UK.

Clarke, K. 1999. Project Corona, sponsored by the National Science Foundation, Santa Barbara: University of California Santa Barbara. Available at <http://www.geog.ucsb.edu/~kclarke/Corona/Corona.html>. Accessed June 27, 2002.

Clarke, K., J. Osleeb, J. Sherry J. Meert, R. Larsson. 1990. The Use of Remote Sensing and Geographic Information Systems in UNICEF's Dracunculiasis (Guinea worm) Eradication Effort. Prev. Vet. Med.11: 229-235.

Climate Prediction Center. 2002. Climate Prediction Center Africa Data Archive. Available at <http://www.cpc.ncep.noaa.gov/products/fews/data.html>. Accessed July 12, 2002.

Cross, E. R., W. W. Newcomb, and C. J. Tucker. 1996. Use of weather data and remote sensing to predict the geographic and seasonal distribution of Phlebotomus papatasi in Southwest Asia. American Journal of Tropical Medicine and Hygiene 54:530-536.

Day, D. A., J. M. Logsdon, and B. Latell (eds.) 1998. Eye in the Sky: The Story of the Corona Spy Satellites, History of Aviation Series, Washington, DC: Smithsonian Institution.

Degregorio, A. and L. J. M. Jansen. 2000. Land Cover Classification System—LCCS, Rome: U.N. Food and Agriculture Organization.

Dickinson, R. E., A. Henderson-Sellers, P. J. Kennedy, and M. F. Wilson. 1986. Biosphere-atmosphere transfer scheme (BATS) for the NCAR community climate model, Boulder, CO: NCAR Technical Note NCAR/TN275+STR.

Earth Satellite. 2002. GeoCover—Global Land Cover Database. Rockville, MD: Earth Satellite, Inc. Available at <http://www.earthsat.com>. Accessed June 5, 2002.

Elvidge, C. D., K. B. Baugh, V. R. Hobson, E. A. Kihn, and H. W. Kroehl. 2002. Detection of Fires and Power Outages At Night Using DMSP-OLS Data. Available at <http://www.ngdc.noaa.gov/dmsp/fires/fires_desc.html>. Accessed June 5, 2002.

Engman, E. T. and R. J. Gurney. 1991. Remote Sensing in Hydrology, New York: Chapman and Hall.

Eva, H. D., A. Glinni, P. Janvier, and C. Blair-Myers. 1998. Vegetation Map of South America at 1:5,000,000. TREES Publications Series D2. Luxembourg, European Commission, EUR 18658 EN.

FAO (U.N. Food and Agriculture Organization). 2002a. Africover: Land cover assessment based on remote sensing for the whole African continent. Available at <http://www.africover.org/General%20over view.htm>. Accessed on June 28, 2002.

FAO. 2002b. Africover Database Gateway. Available at http://www.afri cover.org/adg.htm. Accessed July 12, 2002.

Ferraro, R. R. 1997 SSM/I Derived Global Rainfall Estimates for Climatological Applications. Journal of Geophysical Research 102(16):715-716.

Geist, H. J. and E. F. Lambin. 2002. Proximate Causes and Underlying Driving Forces of Tropical Deforestation. BioScience 52(2): 143-150.

Hassan A. N., L. R. Beck, and and S. Dister. 1998a. Prediction of villages at risk for filariasis transmission in the Nile Delta using remote sensing and geographic information system technologies. J. Egypt. Soc. Parasitol. 28:75-87.

Hassan A. N., S. Dister, and L. Beck. 1998b. Spatial analysis of lymphatic filariasis distribution in the Nile Delta in relation to some environmental variables using geographic information system technology. J. Egypt. Soc. Parasitol. 28:119-131.

Hou, A. Y., S. Zhang, A. da Silva, W. Olson, C. Kummerow, and J. Simpson. 2001. Improving Global Analysis and Short-range Forecasts using Rainfall and Moisture Observations Derived from TRMM and SSM/I Passive Microwave Sensors. Bulletin of the American Meteorological Society 82. pp. 659-680.

IGBP-IHDP. 2002. Land Use and Land Cover Change (LUCC): A Joint IGBP- IHDP Research Project. Available at <http://www.geo.ucl.ac.be/LUCC/scienceplan/scienceplan.html>. Accessed June 5, 2002.

Jensen, J. R. 1996. Introductory Digital Image Processing: A Remote Sensing Perspective. Saddle River: Prentice-Hall, Inc.

Jensen, J. R. 2000. Remote Sensing of the Environment: An Earth Resource Perspective. Saddle River: Prentice-Hall, Inc.

Jensen, J. R. and D. J. Cowen. 1999. Remote Sensing of Urban/Suburban Infrastructure and Socio-Economic Attributes. Photogrammetric Engineering and Remote Sensing. 65:611-622.

Justice, C., E. Vermote, J. Townshend, R. Defries, D. Roy, D. Hall, V. Salomonson, J. Privette, G. Riggs, A. Strahler, W. Lucht, R. Myneni, Y. Knjazihhin, S. Running, R. Nemani, Z. Wan, A. Huete, W. van Leeuwen, R. Wolfe, L. Giglio, J. Muller, P. Lewis, and M. Barnsley. 1998. The Moderate Resolution Imaging Spectroradiometer (MODIS): Land remote sensing for global change research. IEEE Trans. Geosci. Remote Sens. 36:1228-1249.

Kaya, S., T. Pultz, J. Beier, and E. Mushinzimana. 2002. The use of radar remote sensing for identifying environmental factors associated with malaria risk in coastal Kenya. Available at <http://www.pcigeomatics.com/tech-papers/igarss02_kaya_paper.pdf>. Accessed May 31, 2002.

Kimes, D. S., B. N. Holben, C. J. Tucker, and W. W. Newcomb. 1984. Optimal Directional View Angles for Remote-Sensing Missions. International Journal of Remote Sensing 5:887-908.

Kitron, U., L. H. Otieno, L. L. Hungerford, A. Odulaja, W. U. Brigham, O. O. Okello, M. Joselyn, M. M. Mohamed-Ahmed, and E. Cook. 1996. Spatial analysis of the distribution of tsetse flies in the Lambwe Valley, Kenya, using Landsat TM satellite imagery and GIS. J. Animal Ecol. 65:371-380.

Knyazikhin, Y. 2002. MODIS Instrument on NASA's Terra Satellite Improves Global Vegetation Mapping, Makes New Observations Possible. NASA New Archive, Release 01-126, December 20, 2001, Available at <http://earthobservatory.nasa.gov/Newsroom/NasaNews/2001/200112206806.html>. Accessed May 31, 2002.

Lambin, E. F. and H. J. Geist, 2001. Global Land-Use and Land-Cover Change: What Have We Learned So Far? Global Change Newsletter (46):27-30. Available at <http://www.igbp.kva.se>. Accessed June 5, 2002.

Li, Q., R. Ferraro, and N. C. Grody. 1998. Detailed Analysis of the Error Associated with the Rainfall Retrieved by the NOAA/NESDIS SSM/I Rainfall Algorithm: Part 1. Journal of Geophysical Research. 103(11): 419-427.

Linthicum K., C. Bailey, C. Tucker, K. Mitchell, T. Logan, and F. Davies. 1990. Applications of polar-orbiting, meteorological satellite data to detect flooding in Rift Valley Fever virus vector mosquito habitats in Kenya. Med. Vet. Entomol. 4:433-438.

Linthicum, K., C. Bailey, C. Tucker, S. Gordon, T. Logan, and C. Peters. 1994. Man-made ecological alterations of Senegal River basin on Rift Valley Fever transmission. Sistema Terra. Vol 3, pp. 45-47.

Loveland, T., B. Reed, J. Brown, D. Ohlen, J. Zhu, L. Yang, and J. Merchant. 2000. Development of a Global Land Cover Characteristics Database and IGBP DISCover from 1-km AVHRR Data. International Journal of Remote Sensing 21(6-7):1303-1330.

Lulla, K., M. Helfer, and D. Holland. 1994. The NASA Space Shuttle Earth Observations Database for Global Change Science. Pp 335-365 in Remote Sensing and Global Climate Change. R. Vaughan and A. P. Cracknell, eds. Heidelberg: Springer-Verlag.

Malone J., O. Huh, D. Fehler, P. Wilson, D. Wilensky, and R. Holmes. 1994. Temperature data from satellite images and the distribution of schistosomiasis in Egypt. Am. J. Trop. Med. Hyg. 50:714-722.

Malone, J., J.Yilma, J. McCarroll, B. Erko, S. Mukaratirwa, and X. Zhou. 2001. Satellite climatology and the environmental risk of Schistosoma mansoni in Ethiopia and east Africa. Acta Tropica 79:59-72

Mayaux, P., E. Janodet, C. Blair-Myers, and P. Legeay-Janvier. 1997. Vegetation Map of Central Africa at 1:5,000,000. TREES Series D: Thematic output No 1. Ispra, Italy: Space Applications Institute, EC-JRC. EUR 17322 EN.

McDonald, R. A. 1997. Corona, Argon, and Lanyard: A Revolution for U.S. Overhead Reconnaissance. Pp 61-79 in Corona: Between the Sun and the Earth. Bethesda, MD: American Society for Photogrammetry and Remote Sensing.

NASA (National Aeronautics and Space Administration). 2002a. MODIS Land Discipline Web site. Available at <http://modis-land.gsfc.nasa.gov/products/products.asp?ProdFamID=5>. Accessed July 12, 2002.

NASA. 2002b. Rationale for the MODIS Fire and Thermal Anomalies Data Products. Greenbelt, MD: NASA Goddard Space Flight Center. Available at <http://modis-fire.gsfc.nasa.gov/rationale/>. Accessed May 31, 2002.

NASA, 2002c. TRMM-Based 3-hr Real-time Rainfall Analyses, Greenbelt, MD: NASA Goddard Space Flight Center, Available at <http://trmm.gsfc.nasa.gov/publications_dir/precipitation_msg.html>. Accessed May 31, 2002.

NRC (National Research Council). 1995. Earth Observations From Space: History, Promise, and Reality. Washington, DC: National Academy Press.

NOAA (National Oceanic and Atmospheric Administration). 2001a. NOAA's National Environmental Satellite, Data, and Information Service: Economic Value for the Nation. Silver Spring, MD: NOAA. Available at <http://www.nesdis.noaa.gov>. Accessed June 27, 2002.

NOAA. 2001b. National Polar-orbiting Operational Environmental Satellite System (NPOESS). #F90109-99. Washington, DC: NOAA.

Olson, J. S. 1994a. Global Ecosystem Framework: Definitions. USGS EROS Data Center Internal Report. Sioux Falls, SD: USGS.

Olson, J. S. 1994b. Global Ecosystem Framework: Translation Strategy. USGS EROS Data Center Internal Report. Sioux Falls, SD: USGS.

Peebles, C. 1997. The Corona Project: America's First Spy Satellites. Annapolis, MD: Naval Institute Press.

Pope, K., E. Sheffner, K. Linthicum, C. Bailey, T. Logan, E. Kasischke. 1992. Identification of central Kenyan Rift Valley Fever virus vector habitats with Landsat TM and evaluation of their flooding status with Airborne Imaging Radar. Remote Sensing Environment 40:185-196.

Prince, S. D. and C. O. Justice (eds). 1991. Coarse Resolution Remote Sensing of the Sahelian Environment. International Journal of Remote Sensing. 12: 1133-1421.

Robinson, J. A. and C. A. Evans. 2002. Space Station allows remote sensing of Earth to within six meters. EOS, Transactions, American Geophysical Union 83:185-188.

Robinson, T, D. Rogers, and B. Williams. 1997. Mapping tsetse habitat suitability in the common fly belt of southern Africa using multivariate analysis of climate and remotely sensed data. Med. Vet. Entomol.11: 235-245.

Rogers, D. J. 1991. Satellite imagery, tsetse and trypanosomiasis. Prev. Vet. Med.11:201-20.

Rogers, D. and S. Randolph. 1991. Mortality rates and population density of tsetse flies correlated with satellite imagery. Nature 351:739-741.

Rogers, D. and B. Williams. 1993. Monitoring trypanosomiasis in space and time. Parasitology 106 (Suppl):77-92.

Ruffner, K. C. 1995. CORONA: America's First Satellite Program. Washington, DC: Central Intelligence Agency.

Running, S., T. Loveland, L. Pierce, R. Nemani, and E. Hunt. 1995. A Remote Sensing Based Vegetation Classification Logic for Global Land Cover Analysis. Remote Sensing of Environment 51:39-48.

Sellers, P. J., D. Randall, G. Collatz, J. Berry, C. Field, D. Dazlich, C. Zhang, G. Collelo, and L. Bounoua. 1996. A Revised Land Surface Parameterization (SiB2) for Atmospheric GCMs—Part I—Model Formulation. Journal of Climate 9:676-705.

Tappan, G., D. Tyler, M. Wehde, and D. Moore. 1992. Monitoring Rangeland Dynamics in Senegal with Advanced Very High Resolution Radiometer Data. Geocarto International 7(1):87-98.

Tappan, G., A. Hadj, E. Wood, and R. Lietzow. 2000. Use of Argon, Corona, and Landsat Imagery to Assess 30 Years of Land Resource Changes in West-Central Senegal. Photogrammetric Engineering and Remote Sensing 6(6):727-735.

Thomson M., S. Connor, P. Milligan, and S. Flasse. 1997. Mapping malaria risk in Africa: What can satellite data contribute? Parasitology Today 13:313-318.

Townshend, J. and D. Skole. 1995. The global 1 km AVHRR data set. Pp 75-82 in Understanding the Terrestrial Environment, Remote Sensing Data Systems and Networks, P. Mather, ed. Chichester, New York: John Wiley and Sons.

TREES (Tropical Ecosystem Environment Observations by Satellite). 2002. Available at http://www.gvm.sai.jrc.it/Forest/defaultForest.htm. Accessed July 12, 2002.

Tucker, C., H. Dregne, and W. Newcomb. 1991. Expansion and Contraction of the Saharan Desert from 1980 to 1990. Science 253:299-301.

USGS (U.S. Geological Survey). 1998. Declassified Intelligence Satellite Photographs. Fact Sheet 090-96. Washington D.: U.S. Geological Survey.

USGS. 2002a. Global Land Cover Characterization Background. Sioux Falls, SD: EROS Data Center, U.S. Geological Survey. Available at<http://edcdaac.usgs.gov/glcc/background.html>. Accessed May 31, 2002.

USGS. 2002b. EROS Data Center homepage. Available at <http://edc.usgs.gov/>. Accessed July 12, 2002.

U.S. Institute for Peace. 2002. Space Aid: Current and Potential Uses of Satellite Imagery in U.N. Humanitarian Organizations. 24 p. Available at <http://www.usip.org/vdi/vdr/12.html>. Accessed May 31, 2002.

WHO (World Health Organization). 2001. Scaling up the Response to Infectious Diseases: A Way Out of Poverty. Available at <http://www.who.int/infectious-disease-report/2002/framesintro.html>. Accessed May 31, 2002.

ANNEX 6

ANNEX BOX 6-1
High Spatial Resolution Satellite Systems: IKONOS and Quickbird

The IKONOS satellite was launched by Space Imaging, Inc., on September 24, 1999 (<http://www.spaceimaging.com>). The satellite has a linear array remote-sensing system that collects 1 × 1 m (0.45-0.90 μm) panchromatic data and four multispectral visible and near-infrared bands (0.45-0.52 μm; 0.52-0.60 μm; 0.63-0.69 μm; and 0.76-0.90 μm) at 4 × 4 m. IKONOS is in a Sun-synchronous orbit 681 km above Earth. It has cross-track and along-track pointing capability and a nominal swath width of 11 km.

The Quickbird satellite, launched by DigitalGlobe has a linear array remote-sensing system that acquires 61 × 61 cm spatial resolution panchromatic data (0.45-0.90 μm) and four multispectral visible and near-infrared bands (0.45-0.52 μm; 0.52-0.60 μm; 0.63-0.69 μm; 0.76-0.89 μm) at 4 × 4 m. It has a swath width of 20-40 km. Prices for standard panchromatic images are $22.50/km and $25/km for multispectral with a minimum order of 64 km^2 (<http://www.digitalglobe.com>; <http://www.rsi.ca/products/quickbird/news/hir_qbi_news2_041802.htm>).

There are several innovative characteristics associated with these remote-sensing systems.

- Data are obtained using linear arrays to achieve a higher degree of geometric stability in the imagery.
- The orbital platform is not buffeted by atmospheric turbulence, which decreases roll, pitch, and yaw error.
- Data are collected as a continuous swath (e.g., 11 km), reducing the amount of data to be mosaicked.
- Eleven-bit data are superior to previous 8-bit data or film silver halide sensitivity.
- They can obtain overlapping, stereoscopic views of the terrain.
- The sensor is pointable, which increases the probability of obtaining imagery of the area of interest.

ANNEX BOX 6-2
Polar Operational Environmental Satellite Program (POES) and the NOAA Advanced Very High Resolution Radiometer (AVHRR)

The POES program is a cooperative effort between NASA, NOAA, the United Kingdom, and France. The most valuable POES instrument for Agenda 21 issues is the Advanced Very High Resolution Radiometer (AVHRR). AVHRR data are used to study and monitor vegetation conditions in ecosystems, including forests, tundra, and grasslands. Applications include agricultural assessment and land-cover mapping.

AVHRR has a spatial resolution of approximately 1.1 × 1.1 km at the satellite nadir from the nominal orbit altitude of 833 km (517 mi). The AVHRR measures reflected or emitted radiant energy in five spectral bands that vary according to which of nine NOAA AVHRR instruments (numbered by the satellite on which it flies) is being used.

	NOAA-6,8,10	NOAA-7,9,11,12,14,15
Band 1	0.58-0.68 μm	0.58-0.68
Band 2	0.725-1.10	0.725-1.10
Band 3	3.55-3.93	3.55-3.93
Band 4	10.50-11.50	10.30-11.3
Band 5	10.50-11.50	11.50-12.5

Each satellite orbits Earth 14 times daily and has a swath width for each pass of 2,399 km (1,491 mi). NOAA/NESDIS (National Environmental Satellite, Data, and Information Service) in Suitland, Maryland, receives both worldwide recorded and direct readout AVHRR data from the Wallops Island, Virginia, and Gilmore Creek, Alaska, stations. NOAA/NESDIS processes, archives, and reproduces the data.

POES and its continuation, NPOESS (Annex Box 6-3), have the following launch dates:

Sensor Systems	Date
Planned	
METOP–2	Spring 2008
METOP–1	June 2003
NOAA–NPOESS N	January 2008
NOAA–N	December 2003
NOAA – M	June 24 2002
Operational	
NOAA–L (16)	September 21 2000
NOAA–K (15)	May 13 1998
NOAA–J (14)	December 1994

SOURCE: NOAA (2001a).

ANNEX BOX 6-3
National Polar-orbiting Operational Environmental Satellite System (NPOESS)

In 1994 a decision was made to eventually merge the U.S. Department of Defense's (DOD's) Defense Meteorological Satellite Program (DMSP) and the NOAA Polar-orbiting Operational Environmental Satellite (POES) system into a single system: the National Polar-orbiting Operational Environmental Satellite System (NPOESS). The system is jointly managed by NOAA, DOD, and NASA to:

- provide a national operational polar-orbiting, environmental remote-sensing capability;
- achieve savings by converging DOD and NOAA satellite programs;
- incorporate new technologies from NASA; and
- encourage international cooperation.

NPOESS will provide an operational remote-sensing capability from 2008 to 2018. It consists of two satellites in two orbital planes that will replace the DMSP and POES constellations. NPOESS will also include a European Organization for the Exploitation of Meteorological Satellites (EUMETSAT) instrument called the Meteorological Operational Satellite (METOP).

Six NPOESS sensor systems are currently under development (NOAA, 2001b).

- VIIRS (Visible/Infrared Image/Radiometer) collects visible and infrared radiometric data between 0.3 and 14 μm. Data types include atmospheric; clouds; Earth radiation budget; land, water, and sea surface temperature; ocean color; and low light imagery. The VIIRS will combine the radiometric accuracy of the POES AVHRR with the high (0.65 × 0.65 km) spatial resolution of the Operational Linescan System (OLS) flown on DMSP to collect 26 types of environmental data (called environmental data records).
- CMIS (Conical Microwave Imager/Sounder) collects global microwave radiometry and sounding data to obtain information on clouds, sea winds, hurricanes, and rainfall.
- CrIS (Crosstrack Infrared Sounder) measures Earth's radiation to determine the vertical distribution of temperature, moisture, and pressure in the atmosphere.
- GPSOS (Global Positioning System Occultation Sensor) measures the refraction of radiowave signals from the U.S. GPS and Russia's Global Navigation Satellite System to characterize the ionosphere.
- OMPS (Ozone Mapping and Profiler Suite) collects data to permit the calculation of the vertical and horizontal distribution of atmospheric ozone.
- SESS (Space Environment Sensor Suite) collects data on neutral and charged particles, electron and magnetic fields, and optical signatures of aurora.

SOURCE: NOAA (2001a,b).

ANNEX BOX 6-4
The Earth Observing System Terra Satellite: Moderate Resolution Imaging Spectroradiometer (MODIS)

NASA's Terra satellite is the flagship of the Earth Observing System (EOS). This research satellite was launched on December 18, 1999, into a 705-km Sun-synchronous orbit. It contains five remote-sensing systems: the Advanced Spaceborne Thermal Emission and Reflection Radiometer (ASTER), the Clouds and the Earth's Radiant Energy System (CERES), the Multi-angle Imaging SpectroRadiometer (MISR), the Moderate Resolution Imaging Spectroradiometer (MODIS), and the Measurements of Pollution in the Troposphere (MOPITT).

MODIS began collecting science data on February 24, 2000. It views the entire surface of Earth every one to two days. It has a field of view of $\pm$ 55° off-nadir, which yields a large swath width of 2,330 km. MODIS obtains high radiometric resolution images (12 bit) of daylight-reflected solar radiation and day/night thermal emission over all regions of the globe. MODIS's scanning, imaging radiometer collects data in 36 co-registered spectral bands: 20 bands from 0.4-3 μm and 16 bands from 3-15 μm. MODIS has one of the most comprehensive calibration systems ever flown on a remote-sensing instrument.

MODIS's relatively coarse spatial resolution ranges from 250 × 250 m (bands 1-2) to 500 × 500 m (bands 3-7) and 1 × 1 km (bands 8-36). Consequently MODIS is valuable for regional Earth-resource analyses, especially those dealing with vegetation characteristics and water quality. MODIS data can be used to compute an Enhanced Vegetation Index (EVI), Leaf Area Index (LAI), Fraction of Photosynthetically Active Radiation (FPAR), and surface temperature. There are approximately 40 MODIS data products. These products arise from MODIS Scientific Algorithm Theoretical Basis Documents (ATBD's) that convert the radiances received by the instrument into geophysical quantities. The following list summarizes some of the major ATBDs associated with MODIS land applications.

MODIS Land Algorithm Theoretical Basis Documents (ATBDs)

Document	Name
ATBD-MOD-08	Atmospheric Correction Algorithm
ATBD-MOD-09	Surface Reflectance: Reflectances BRDF/Albedo
ATBD-MOD-10	Snow and Sea Ice Mapping Algorithm
ATBD-MOD-11	Land Temperature and Emissivity
ATBD-MOD-12	Land Cover
ATBD-MOD-13	Vegetation Indexes
ATBD-MOD-14	Thermal Anomalies, Fires and Biomass Burning
ATBD-MOD-15	LAI (Leaf Area Index) and FPAR (Fraction of Photosynthetically Active Radiation)
ATBD-MOD-16	PSN (daily photosynthesis) and ANPP (Annual Net Primary Production)
ATBD-MOD-29	Enhanced Land Cover and Land Cover Change

ANNEX BOX 6-5
Calculation of Vegetation Indexes from AVHRR data

Vegetation indexes from AVHRR data are based on mathematical modeling of spectral reflectance measurement in various parts of the electromagnetic spectrum. Most indexes make maximum use of the fact that vegetation absorbs much of the incident blue and red radiant energy and reflects much of the incident near-infrared radiant energy (Jensen, 2000). This inverse relationship is based on the physiological structure of healthy living vegetation and how light interacts with the vegetative matter. For example, the Normalized Difference Vegetation Index (NDVI) is calculated using the near-infrared (NIR) and visible (VIS) bands and the following relationship (Kimes, et al., 1984):

$$NDVI = (NIR - VIS)/(NIR + VIS).$$

The magnitude of the NDVI is related to the level of photosynthetic activity in the vegetation. NDVI is a non-linear function that varies between −1 and +1 (undefined when NIR and VIS are zero). Values of NDVI for vegetated land generally range from about 0.1 to 0.7, with values greater than 0.5 indicating dense vegetation.

ANNEX BOX 6-6
Measuring Hydrologic Variables by Remote-Sensing

Hydrologic variables such as precipitation and soil moisture can be detected using passive microwave sensors mounted on satellites or aircraft. Earth materials do not emit a tremendous amount of passive microwave energy, however a suite of radiometers have been developed that record subtle levels of passive microwave energy (Engman and Gurney, 1991).

A scanning passive microwave radiometer collects data across-track as the aircraft or satellite moves forward. The result is a matrix of brightness temperature values that form a passive microwave image. Passive microwave radiometers generally record energy at frequencies between 1 and 200 GHz (at wavelengths of 0.15 to 30 cm). The most commonly used frequencies (channels) are centered at 1, 4, 6, 10, 18, 21, 37, 50, 85, 157, and 183 GHz. The recorded bandwidths (range of frequencies) are usually fairly broad so that enough passive microwave energy is available to be recorded by the antenna. For the same reason the spatial resolution of passive microwave radiometers is usually very large. Aircraft sensors flying closer to the ground may have spatial resolutions measured in meters while most satellite passive microwave scanning radiometers have a spatial resolution measured in kilometers (Jensen, 2000).

Rainfall Detection from the Defense Meteorological Satellite Program

The Special Sensor Microwave/Imager (SSM/I) was one of the first passive microwave sensors to obtain global passive microwave information. Beginning in 1987 it flew onboard the Defense Meteorological Satellite Program (DMSP) satellites. Annex Table 6-1 identifies the DMSP satellites that have carried the SSM/I sensor and the dates of data collection. The SSM/I is a four-channel, polarized passive microwave radiometer that measures atmospheric, ocean and terrain microwave brightness temperatures at frequencies of 19.35, 22.23, 37.0, and 85.5 GHz. The SSM/I rotates continuously about an axis parallel to the local spacecraft vertical and measures the upwelling scene brightness temperatures. The swath is approximately 1,400 km.

The area of the image for each channel (its "footprint") varies with channel energy, position in the scan, along scan or along-track direction, and altitude of the satellite. The 85-GHz footprint is the smallest with a 13 × 15 km and the 19-GHz footprint is the largest at 43 × 69 km. DMSP satellites are in a Sun-synchronous, low altitude polar orbit. The orbital period is 101 minutes and the nominal altitude is 830 km. The data are transmitted to NOAA/NESDIS in Suitland, Maryland.

NOAA personnel and others developed SSM/I rainfall algorithms that use the 85.5 GHz channel to detect the scattering of upwelling radiation by precipitation-size ice particles within the rain layer. Rain rates can be derived indirectly, based on the relationship between the amount of ice in the rain layer and the actual rainfall on the surface. In addition, a scattering-based global land rainfall algorithm has been developed. Monthly rainfall at 100 × 100-km and 250 × 250-km grids have been produced for the period from July 1987 to the present (Ferraro, 1997; Li et al., 1998).

Rainfall Detection from the Tropical Rainfall Measurement Mission

The TRMM Microwave Imager (TMI) is a passive microwave sensor designed to provide quantitative rainfall information over a 487-mile (780-km) swath. It is based on the design of the SSM/I and measures the intensity of radiation at five frequencies: 10.7 (45-km spatial resolution), 19.4, 21.3, 37, and 85.5 GHz (5-km spatial resolution). Dual polarization at four frequencies provides nine channels. The new 10.7-GHz frequency provides a more linear response for the high rainfall rates common in tropical rainfall (Jensen, 2000).

Calculating the rainfall rates from both the SSM/I and TMI sensors requires complicated calculations because water bodies such as oceans and lakes emit only about one-half the energy specified by Planck's radiation law at microwave frequencies. Therefore, they appear to have only about half their actual temperature at the surface and appear very "cold" to a passive microwave radiometer. Fortunately, raindrops appear to have a temperature that equals their real temperature and appear "warm" or bright to a passive microwave radiometer. The more raindrops, the warmer the whole scene appears. Research over the last three decades has made it possible to obtain relatively accurate rainfall rates based on the temperature of the passive microwave scene.

Land is different from oceans because it emits about 90 percent of its real temperature at microwave frequencies. This reduces the contrast between the rain droplets and the land. Fortunately, the high-frequency microwaves (85.5 GHz) are strongly scattered by ice present in many raining clouds. This reduces the microwave signal of the rain at the satellite and provides a contrast with the warm land background, allowing accurate rainfall rates to be computed over land as well as water.

Rainfall Detection from NASA's Aqua Satellite

The Advanced Microwave Scanning Radiometer-EOS (AMSR-E) is flown on NASA's Aqua satellite in a polar, Sun-synchronous orbit. It was launched on May 4, 2002. The 12-channel passive microwave radiometer measures frequencies at 6.9, 10.7, 18.7, 23.8, 36.5, and 89 (HV polarization) and 50.3 and 52.8 (VV polarization). It has a 7-km field of view at 89 GHz and 60-km field of view at 6.9 GHz and a 1,600-km swath width. The AMSR measures total water-vapor content, total liquid-water content, precipitation, snow-water equivalent, soil moisture (using the 6.9- and 10.7-GHz frequencies), sea-surface temperature (SST), sea-surface wind speed, and sea-ice extent. The specifications of the sensors are subject to change.

ANNEX TABLE 6-1 Defense Meteorological Satellite Program Operational Line Scan and SSM/I Data Available for Distribution from the Various Satellites (e.g., F10 to F14)

Sensor	F10	F11	F12	F13	F14
OLS	4/12/92-2/8/95[a]	4/12/92-4/22/95[a]	9/25/94-present	4/24/95-present	4/28/97-present
SSM/I	4/12/92-11/14/97	4/12/92-4/22/95 4/21/97-present	None[a]	4/24/95-present	4/28/97-present

[a]Data stopped due to sensor problems.

7

GIS-Based Decision-Support Systems in Africa

INTRODUCTION

The previous chapters discuss the importance of compiling fundamental data and their integration within spatial data infrastructures. This is a necessary first step toward informed decision-making, for without accurate, reliable, and interdisciplinary data from a variety of sources it will be difficult to evaluate the many facets of sustainable development (Table 1-1). When observations are made on a routine basis, as can be done with remote-sensing, and when various sources and types of data are accessible and can be integrated into a GIS, the basic infrastructure exists for making informed decisions. This chapter discusses the concepts for developing and using geographic data within spatial decision-support systems, examples of applications of these systems, impediments to their use, and opportunities for enhancing their application to challenges of sustainable development.

The Value of Spatial Decision-Support Systems

The action of accumulating data is in itself insufficient to assess and manage the complex process of sustainable development and its broad implications for the environmental, economic, health, and social issues that confront policy-makers and citizens. Capacity is needed to evaluate natural resources and environmental change in a strategic planning context, primarily through analysis or forecasting of change with models that are sensitive to policy options.

Forecasting tools are common in economic analyses, and planners and policy-makers routinely use estimates from economic and demographic analyses and forecasts to evaluate alternatives and inform their decisions. Yet, geographic information activities in Africa have tended to focus on making maps or collecting observations (EIS-Africa, 2001). An analytical framework in spatial decision-support systems is needed for environment and natural resource analysis and forecasting.

Spatial decision-support systems are interactive and computer-based (Malczewski, 1997). They help prioritize dataset development and information gathering and put geographic information in a decision-making context. These systems are used to set data and information requirements in terms of time, scale, accuracy, and methods, and make it possible to link spatial decision-support efforts with other planning efforts, such as economic development, transportation planning, or poverty reduction programs.

A spatial decision-support system allows a decision-maker to (1) build relationships, both spatial and process-based, between different types of data, (2) merge multiple data layers into synthetic information, (3) weigh outcomes from potentially competing alternatives, and (4) forecast. To do this a spatial decision-support system uses three basic elements: (1) data (e.g., see Chapters 5 and 6), (2) known relationships between data, and (3) analysis functions and models to synthesize relationships or to test scenarios of different policy or decision-making alternatives.

Spatial decision-support tools, such as GIS, are important facilitators of the use of geographic data for sustainable development. As stated in *Our Common Journey* (NRC, 1999a),

> Ultimately, success in achieving a sustainable transition will be determined not by the possession of knowledge, but by using it, and using it intelligently in setting goals, providing needed indicators and incentives, carefully examining alternatives, establishing effective institutions, and, most generally, encouraging good decisions and taking appropriate actions.

To support good decisions, spatial decision-support systems that include GIS often are employed.

Geographic Information Systems

Geographic information systems provide an excellent medium for data integration and a basis for a spatial decision-support system (Cowen, 1988). A GIS supports decision-making by providing ways to examine and choose among alternative solutions, and takes decision-makers beyond the point of simply possessing data, information, and knowledge.

The concept of a decision-support system dates to the late 1950s (e.g., Simon, 1977), and systems using geographic data emerged in the last 10 years and subsequently grew dramatically (Densham and Goodchild, 1989; NCGIA, 1990, 1996). By 2000 over $900 million had been spent on GIS software, and GIS-related services had generated $7 billion in revenue. The annual growth rate of the GIS industry is now 15 to 20 percent, and there are over 2 million GIS users worldwide (Daratech, 2001).

Still, little of the growth in the GIS industry has taken place in Africa. Increased use of GIS in African countries depends on effective demand for geographic information and tools, and on technology cooperation with other countries. Lessons learned from past efforts to transfer technology from developed to developing countries are (Schmidheiny, 1998)

- technology ought to be appropriate to serve user needs.
- people should be educated and trained in the use of the technology.
- technology should suit local conditions (e.g., climate, energy availability, customs).
- technology should be transferred over the long term.

GIS technology is available at a variety of technology levels, or scales of implementation, from advanced systems using considerable computational power and large datasets to use of paper maps and GPS. The level of technology should be appropriate to its intended use. Hence, this report encourages the use of geographic information—not GIS technologies per se—to support decision-making, regardless of its level of technological advancement.

DECISION-MAKING AND GEOGRAPHIC INFORMATION

Decision-makers in African countries need data and tools to monitor and assess natural resource inventories and environmental and social change. These data and tools are also needed to predict scenarios (e.g., trends and needs for land and food), determine critical information needs, evaluate data quality, and identify data gaps. Entities ranging from governments to NGOs to farmers can use information from decision-support systems to reduce the impact of global change on human well-being and the environment. Needs and priorities vary among these entities. Therefore, decisions about sustainable development often involve compromises and trade-offs (e.g., setting aside land for wildlife protection versus land for farming, or deciding how much water from a river should be diverted to farming as opposed to industry or housing), and competing demands complicate the decision-making process.

From GIS to Decision-Support

A GIS aids the decision-making process by integrating and displaying data in an understandable form. Furthermore, a GIS is used to analyze relationships among different kinds of data (e.g., environmental and health data). The fundamental analytical functions of a GIS-based spatial decision-support system include (1) query analysis, (2) proximity or buffer analysis, (3) overlay analysis, (4) neighborhood analysis, (5) network analysis, and (6) modeling (Box 7-1). Various combinations of these functions are commonly used during the geographic data analysis process.

GIS is not an end in itself, however, but provides a valuable foundation for further analysis. A spatial decision-support system can be based on the primary functions of a GIS, but these basic functions need enhancements for analysis and modeling. For instance, for food-security analysis it is possible to link a GIS to a model that predicts grain yields from a range of spatial input data, such as soils, climate, and topography. This model can be linked to economic and demographic models showing where people live and the grain demand from these settlements. The combination of basic data, yield modeling, and human demand and location analysis provides a way to evaluate food security. Hence, using a spatial decision-support system is not simply a descriptive exercise. The desired outcome is not how the world looks, but instead how the world works.

A critical feature of a spatial decision-support system is its emphasis on linking data with analysis tools. Some analyses use spatial analysis functions often referred to as GIS modeling in which several data layers are merged to create a new synthetic layer. This is often the approach for risk assessment. For example, various habitat, human population, and climate data layers can be merged to provide a vector-borne disease risk map as a product (e.g., Chapter 6). This product can be updated rapidly and often is a means to marshal scarce resources.

Spatial decision-support systems also can involve numerical models, including forecast models that evaluate through simulation in map form various alternative scenarios based on different policy options. This type of decision-support usually is deployed for planning purposes. It can be useful as a way to integrate multiple planning objectives, or competing options for the use of a specific natural resource, such as land, or particular location, such as a watershed.

Lastly, a spatial decision-support system is indeed for support to decision-making—it does not make decisions by

BOX 7-1
Analytical Functions of a GIS-Based Decision-Support System

Query Analysis. A query is a question asked of the decision-support system. For example, a GIS could be asked to show all primary schools, water wells, or markets within a specified geographic area. Queries with more specific conditions might include:

- Where are all the lateritic soils?
- Where are croplands that are at risk from high erosion?
- Where are all the paved roads in an administrative area?

Proximity or Buffer Analysis. Buffer analysis is a geographically or temporally constrained version of query analysis. The GIS creates a buffer or boundary of a specified distance (measured in units of length or time) around an object represented as a point, line, or polygon. The buffer is then used to constrain the queries to within a specified distance.

The types of questions that might be asked using buffer analysis include:

- Where are all the people that fall within a specified distance of a clinic (a point)?
- Or within a specified distance of a river (line) (e.g., to determine a region's dependence on a particular water system)?
- Or within a specific distance of a city boundary (polygon)?

Overlay Analysis. This analysis involves the "electronic stacking" of spatial data (e.g., human population, land cover, soils, hydrology) as layers on "top" of each other so that the geographic position within each layer is precisely registered to all the other data layers in the database (Figure 1-2).

Queries that might be addressed using overlay analysis include:

- Show all locations where a particular vegetation type is growing on a specified soil type (vegetation layer and soil-type layer).
- Determine the distribution of people exposed to disease-vector (e.g., mosquito) habitats to show populations at greatest risk of health problems (population layer, hydrography layer, elevation layer, health-center layer).
- Identify those areas where agricultural production may be most feasible and provide the greatest benefits using an area's different soil types overlaid by vegetation types and population density.

Neighborhood Analysis. This determines the characteristics of features that are in close proximity (neighbor) to an object or an area of interest. A moving window is used; for example, a window might be systematically moved across a data layer to determine the statistical characteristics of the pixels within the window such as the average elevation if the data layer were topography.

Network or Connectivity Analysis. This is used on vector-based datasets to determine such network characteristics as the shortest route to a clinic.

Modeling. Because data in a GIS can be accessed, transformed, and manipulated interactively, a GIS can serve as a testing area for analyzing processes, analyzing the results of trends, or projecting the possible results of decisions. The use of GIS allows non-destructive experimentation and manipulation of the environment and other factors. Changes in the geographic characteristics of features such as size or shape can be modeled over time. For example, land-use changes, such as changing farming practices can be modeled to predict per-hectare loss of soil over time.

itself. Human inputs must be recognized, particularly in the complex arena of sustainable development. A spatial decision-support system is not simply a turn-key process by which data are ingested and buttons are pushed. Humans select the questions posed and which data and analyses to use. Humans also decide how to disseminate results (e.g., which medium will most effectively convey the information to those who need it) and encourage open participation in the process.

Decision-Makers and Data Users

The data and technology needs of decision-makers vary with the types of users. Policy-makers at the national level need different information than the residents of a town or village affected by the decision. Although good decision-making involves all the people who are likely to be affected by a decision, individuals and organizations play different roles in the decision process. People and groups are not always comfortable or familiar with direct responsibility for decision-making (e.g., Eastman, 1999, 2001a,b). For example, to decide where to build a road, policy-makers will want to know all options and be able to calculate the outcomes of alternatives. Local people, on the other hand, may want to express their priorities and views, yet defer the actual decision to others. Stakeholders in a decision, that is, those most involved in and affected by the decision, can still be indirectly involved in the decision-making process.

Technology must be useful and appropriate to its users. For example, African scientists require advanced GIS and modeling capabilities. New technology might be introduced to scientists and technical users through partnerships with their counterparts in developed nations. Non-technical users, such as policy-makers or community leaders, on the other hand, might find printed material such as brochures, satellite imagery, sample data, or maps more useful. Projects operate at both ends of this technology continuum in Africa.

Data for Decision-Support Systems

The first requirement for implementing a spatial decision-support system is access to data (Chapter 4). Ideally, decision-support systems use distributed GISs so that users can obtain data relevant to their needs, such as framework data and other thematic data (Chapters 5 and 6). A geolibrary is an example of an open distributed system that combines the idea of a traditional library with the resources of the Internet. Geolibraries make geographic data available to those with access to a computer and the Internet (NRC, 1999b).

Distributed geolibraries are global in reach and are part of the concept of the national and global spatial data infrastructures (Chapter 4). Data-sharing, necessary for a distributed system, often is inhibited by a lack of precedent and protocols for sharing among government agencies (EIS-Africa, 2001) and other entities. Fortunately, a number of African countries are creating protocols for data-sharing (e.g., Ghana, Mozambique, Senegal, Uganda, and Zimbabwe) as part of their participation in EIS-Africa (Chapter 4).

Types of Decisions

Decisions are on a continuum ranging from structured to unstructured. Structured decisions can be solved by computers. They require only manipulation of information and mathematical computations. Unstructured decisions involve human judgments, such as assessing risk or priorities, or human values like determining what is just or fair. These kinds of decisions cannot be made by a computer. Most decision problems fall somewhere between these two extremes and are called semi-structured decisions. Spatial decision-support systems provide computation and analytic power for structured decisions and model alternative solutions for human consideration. Through this process a semi-structured decision is made.

The process of formalizing the development of a decision-support system and related organizational requirements is referred to as "managed decision support." J. R. Eastman (Clark University, personal communication, 2001) describes a managed decision process for applying geographic information to sustainable development that has three types of decisions: (1) resource allocation decisions, (2) resource status decisions, and (3) policy decisions.

Resource Allocation Decisions

The first step in resource allocation decision-making is the standardization of information including units of measurement and data accuracy. Assuring the accuracy of data is an important next step that adds cost and requires trained staff to collect field data for validation (ground-truthing).

Resource allocation decisions often involve tradeoffs or assessment of risk. These decisions are value questions as well as technical questions. Support for this type of decision includes (1) aggregation and weighting that enable mapping of priority and risk areas and (2) appraisal of options by modeling. Resource allocation decisions are also characterized by multiple objectives, such as food production and disease control. A decision-support system for questions about farming and disease would incorporate data on land cover, population distribution, hydrology, and other factors necessary to analyze and map the risk of disease and food insecurity. A GIS for this purpose would also allow users to analyze tradeoffs and weigh outcomes of alternative plans.

Resource Status Decisions

Resource status decisions involve merging routine observations of the status of a resource (e.g., timber, cattle, or fuel) with policy planning and management issues. Decision support requires routine, repeated collection of data on the desired parameters. There is a tendency in GIS technology transfer to focus on static factors, such as political units, elevation and slope, and other framework data sets (Chapter 5). Although these datasets are required for construction of base maps, they do not by themselves support decisions that require assessment of status and changing conditions. Analytical tools are needed to highlight changes.

FEWS (Chapters 3 and 6) is among the most successful demonstrators of resource status decision support in Africa. In FEWS, data come from routine updates of vegetation condition, through the Normalized Difference Vegetation Index (NDVI). The changes in NDVI provide an indication of moisture conditions that gives advance warning of drought. FEWS provides rapid decision support at localized scales using direct observations. Similar observations can be made for drought and fire risk; for example, the Miombo Network (Chapter 3) and programs of the MODIS land science team (Chapter 6).

Although local applications of geographic information are important, environmental effects on a large scale must also be monitored. A continental-scale application of condition and status assessment comes from programs that are developing early warning capability for El Niño[1] prediction. El Niño and land use are linked. Understanding El Niño has important implications for food productivity. Systems that couple large-scale advance warning with current regional or local conditions provide decision-makers with more time to respond.

Policy Decisions

A challenge for decision support is to assess the future impact of implementing policy options. For example, analysis of future land-cover changes as a result of future land-use distribution requires analytical capabilities that are time and

[1]An episodic global weather phenomenon driven by conditions in the western Pacific Ocean.

space sensitive. Many predictive land-use and cover-change models are simple extrapolations from trends in variables such as estimates of population growth. Policy decision-making requires the analytic and modeling capabilities of GIS.

Many policy decisions require spatial data and analysis of overlapping sets of data layers. Necessary data analysis includes computer functions that merge various layers to produce a synthetic layer. For example, soil erosion risk maps are produced by merging land use, slope, and soil information.

EXAMPLES OF DECISION-SUPPORT SYSTEMS IN AFRICA

Geographic information technologies are used in African countries and elsewhere, but are rarely used in routine support of policy-making, natural resource management, or planning (e.g., EIS-Africa, 2001). Five examples illustrating different aspects of implementation of decision-support systems in Africa at scales ranging from the continental to the community level follow:

1. A Continent-wide Application—Mapping Malaria Risk in Africa (MARA) Project

The Mapping Malaria Risk in Africa (MARA) project maps malaria risk using in situ data on malaria occurrence in combination with spatial modeling to predict the geographic distribution, seasonality, and endemicity (peculiarity to a locality or region) of the disease. The project uses the GIS to evaluate location and risk and to disseminate information to national and international decision-makers.

The MARA project is a federated network of scientists throughout Africa who are mapping malaria risk at the district level. Five regional data collectors are responsible for obtaining malaria datasets from neighboring countries. Stratified risk maps of the type and severity of malaria transmission are produced from geographic data on demography, climate, elevation, ecological zone, vector distribution, and malaria endemicity.

MARA uses continent-wide datasets (e.g., land cover, elevation, biotype) in addition to local precipitation and temperature data. This combination of information from local and continental-level sources presents challenges for accuracy and compatibility, as does the need to organize data on an administrative district level. Health authorities in each country rank each district in terms of the severity of the malaria risk. This derived product of risk-ranking assessment is then mapped in the GIS and provided to health officials. MARA provided the first continental maps of malaria distribution and the first quantitative "burden of disease" estimates. Maps produced by MARA are widely used for planning, intervention, and prevention. In the committee's opinion, this simple decision-support system is effective because it relies on low-cost data, operates on a routine basis, and involves a broad network of African institutions and scientists.

2. A Continent-wide Application—The Famine Early Warning System Network

FEWS NET is a network of 17 African countries working with partners to address food security issues (Chapter 3). FEWS NET operates on the principle that gradually unfolding natural disasters influencing food security give decision-makers time to prepare and take preventive action.

A range of data and information sources is used by FEWS NET including continent-wide, 10-day NDVI (Chapter 6) and rainfall estimates from NOAA and European satellites, ground-based meteorological data, data on crop and rangeland conditions, commodity pricing data and agricultural production data (Chopak, 2000). FEWS NET handles a large volume of data, and has developed automated processing and analysis tools for routine operations.

Food-security analysis is broken into five assessment activities: a start-of-season assessment, a preliminary crop forecast, an annual food balance sheet, a harvest assessment, and a current vulnerability assessment (Chopak, 2000). After the analysis stage, information is disseminated to decision-makers. Monthly updates for all member countries are posted on the FEWS NET website (FEWS NET, 2002). Additionally, memoranda are issued to warn of developing food security issues (e.g., Box 7-2).

The USGS provides technical support to FEWS NET partners in the use of remote-sensing and GIS and develops data-processing and analysis tools. Additionally, USGS assists with data archival and dissemination.

3. A Regional Network Application—Fire Detection and Response in the Miombo Woodlands

The Miombo Network was formed to create a regional network for environmental research on the dominant biome[2] of southern Africa, the Miombo Woodlands (Chapter 3). It is an informal network of scientists funded through grants and contracts from donor and science agencies.

One of the important environmental threats in the region is fire. The Miombo Network has developed a remote-sensing approach to fire detection and mapping using the MODIS sensor onboard NASA's Terra satellite (Chapters 5 and 6). MODIS data are retrieved and provided on an Internet-based interactive GIS, which shows the location of fires and the underlying vegetation cover.

Remotely sensed vegetation maps are incorporated into the GIS and merged with in situ data to derive fuel loads, which are then mapped and merged with fire location. The

[2]A biome is an ecological formation including both plants and animals. Traditionally biomes are identified in terms of their characteristic vegetation form.

BOX 7-2
Sample Early Warning Memorandum of an Evolving Food Security Crisis

August 15, 1999

RE: LIVESTOCK SITUATION IN NORTHERN BOTSWANA WORSENING

The livestock situation in northern Botswana continues to deteriorate as a result of spreading disease and the late delivery of veterinary supplies. As a result of too much rainfall during the past season, the conditions were existent for a livestock disease outbreak. Since April over eighty percent of cattle have died. Although animal deaths are not unusual after heavy rains experienced this year in this part of the country, the lack of available drugs have exacerbated the already difficult situation. The result is that the productive capacity of farm households in this area will have difficulty in land preparation in the coming agricultural season.

SOURCE: Chopak (2000).

integration of fire location, area burned, vegetation type, and fuel loads enables a near-real-time prediction of fire risk and active fire. Decision-makers use these maps, which are available on the Internet, to target limited resources to areas of greatest need. In 1997 the Miombo Network produced a CD that includes a Web-based interface, GIS software, and geographic data archived in a standard format (Arcinfo export format). This CD was distributed widely and freely in Africa to overcome data access problems.

In the committee's opinion, the Miombo Network is successful and self-sustaining because it is local need-driven, African-directed, and relevant to the missions of its participating agencies.

4. A Regional Example—The Livestock Early Warning System

Food security and famine in East Africa are related to weather variation, expanding human populations, political instability, and changing patterns of land use and land tenure. The Livestock Early Warning System (LEWS) project, which is funded by USAID and is being implemented by Texas A&M University, demonstrates the application of integrated remotely sensed weather data, point-based biophysical modeling, and geographic data on animal and vegetation distribution to serve decision-makers concerned with the welfare of pastoral communities in East Africa.

Typically, early warning systems (e.g., FEWS [Chapter 6]) provide predictive data on rainfall and vegetation condition, whereas the on-ground monitoring programs of markets, human conditions, and animal herd situations provide a "post-effect" appraisal system. However, many of the problems affecting livestock (e.g., weight loss, loss of condition) occur before the response is visible, irrespective of personal experience. This problem necessitates an early warning system that works at the local level (LEWS, 2001).

In response to this challenge LEWS applies technologies capable of (1) predicting the current nutritional status of free-ranging animals, (2) assessing the impact of weather on forage supply and crop production, and (3) linking these data with local, household-level data. Currently, there are six operational monitoring zones in East Africa (southern Ethiopia, northern Kenya, southern Kenya, central and southwestern Uganda, northern Tanzania, and central Tanzania), with two other zones under development in northeastern Ethiopia and northeastern Uganda. Each zone is comprised of 30 LEWS monitoring points. A network of households in each zone provides monthly fecal samples from cattle, sheep, and goats. Nutritional well-being of free-ranging livestock is then assessed in laboratories through fecal profiling using a technology called near-infrared spectroscopy. The fecal profiles are geographically referenced and integrated into the GIS, along with livestock population surveys, continuous 10-day weather data sets, and vegetation data (NDVI [Chapter 6]) from the FEWS program. This combination of data sources provides a foundation dataset for a linked series of predictive models involving

- a grazing land production model;
- a livestock nutritional balance analysis model;
- a mixed farming crop model; and
- a modified El Niño/Southern Oscillation model calibrated for East Africa.

Every 10 days a small data file containing the output of the predictive models is packaged for broadcasting to teams equipped with World Space Satellite radios (Chapter 4) linked to laptop computers. This process is managed by the Association for Strengthening Agricultural Research in Eastern and Central Africa's (ASARECA) Crisis Mitigation Office in Nairobi and distributed on the African Learning Channel with help from the Arid Lands Information Network.

Over the past two years LEWS has cultivated partnerships with the Intergovernmental Authority on Development Drought Monitoring Center and the Regional Center for Mapping of Resources for Development (RCMRD [Chapters 3 and 8]), both in Nairobi. Currently LEWS is in the process of improving computer server and analytical capacity at these organizations and providing training in the use of Arcview GIS software. Texas A&M University's Center for Natural Resource Information Technology provides staffing and server capacity to support the system.

A final component of the LEWS activity is community outreach to pastoralists. LEWS is testing early warning communication techniques in pastoral villages, with the aim of more effectively influencing decision-making at the local level.

5. A Community-Based Example—Natural Resource Management in Namibia

The quality of local involvement in natural resource decision support may be more important than the level of technical sophistication. In Namibia communities are experimenting with the use of paper maps generated from a GIS to manage natural resources at a community level (Figure 7-1). This work demonstrates the creation of long-term self-sustaining applications of geographic information to sustainable development.

The principal applications of geographic data in the Community Based Natural Resource Management (CBNRM) program are for mapping conservancy boundaries, land-use planning, monitoring, and communication.

- Accurately defined boundaries are a legal requirement for registration in the conservancy. Neighbors must agree upon boundaries. Although these discussions can be contentious, displaying the proposed boundaries in a GIS map has proven to be an effective decision-support tool.
- Land-use decisions include evaluating competing activities within conservancies such as farming, settlement, mining, tourism, and wildlife management. To aid in this process staff at the central office in Windhoek gather and integrate pertinent geographic data within a GIS and disseminate the resultant paper map products to the conservancies. Through an iterative process of participatory mapping, members of the conservancies converge on solutions.
- Geographic information assists in annual game counts, monitoring poaching or problem animals, assessing drought risk, and monitoring trends in animal populations. Maps are central in all three stages of the monitoring process: planning, implementing, and reporting.
- Communication within conservancies and externally to neighbors, partners, donors, and potential investors is aided by geographic information. Maps with icons (rather than words or numbers) are effective in overcoming communication challenges. And well-organized land-use plans attract potential investors who use maps of game distribution and land-use zones in their deliberations about the placement of tourist lodges that provide income and jobs for local people.

FIGURE 7-1 Paper maps used in decision support in Namibia (Source: Jo Tagg, Namibia Nature Foundation).

The program operates on a needs-driven approach, and promotes sharing of data and tools. Sharing has resulted in considerable cost savings, and is promoted by not charging for data; adopting standardized software, data formats, and file directory structure; and developing a metadata database. The needs-driven approach has built credibility with field users, led to a strong feeling of ownership by rural people and field-based support staff, fostered a culture of sensitivity to community needs among technical institutions that are partners in the program, generated trust and a common vision among partners (communities, government, donors), and built a critical mass to enhance sustainability of the program (Jo Tagg and Greg Stuart-Hill, Namibia Nature Foundation, personal communication, 2002).

IMPEDIMENTS TO IMPLEMENTING SPATIAL DECISION-SUPPORT SYSTEMS IN AFRICA

EIS-Africa (2001) concluded from a review of information initiatives in Ghana, Mozambique, Senegal, Uganda, and Zimbabwe that few application-oriented examples demonstrated advanced analysis of geographic information. Information systems were still insignificant in environmental decision-making. The following reasons were cited:

- the decision-support process has not been planned, and clear objectives, goals, and responsibilities of local agencies have not been set;
- projects are orientated toward data production and updating rather than usage or application (most organizations do not consider analysis to be part of their mission);
- there is a focus on technical issues instead of data management in support of the decision-making process;
- many databases are still under construction—hence many organizations have not had time to consider analysis;
- poor communication between technical staff and those involved in environmental science;
- lack of inclusion of universities in the environmental network in several of the countries (reducing the chance for ongoing research projects to be a driving force for analysis and data combination);
- lack of demand and user awareness; and
- difficulty measuring the impact of technological advances on the decision-making process.

The focus on developing data instead of the analytical environment for using data in the decision-making process partly results from limitations on data accessibility. However, there are also organizational barriers caused by a focus on applied analysis rather than basic research. Research investments have the potential to advance understanding of relationships that tie data together. For example the relationship between soil texture, slope, rain intensity, and other factors that determine soil erosion is embodied in the Universal Soil Loss Equation, which itself must be locally calibrated. This equation is a powerful analytical tool, and was an outcome of basic research. It remains to be tested and adapted for use in many parts of Africa.

Deploying a spatial decision-support system requires field research, case studies, and pilot projects. Development assistance investments targeted to research-based programs would promote a shift from descriptive to process-based analysis in spatial decision-support systems. There are some examples where these long-term investments can work. USAID's Cooperative Research Support Program (Chapter 3) recognizes the importance of research to development assistance programs and has promoted collaborative research between U.S. land grant universities and African organizations, mostly in agriculture and natural resource management. With the increasing availability of geographic data and decision-support tools, there is an opportunity for these programs to emphasize the spatial aspects of the research.

OPPORTUNITIES FOR ENHANCING DECISION SUPPORT IN AFRICA

This section draws from lessons learned in examples from the previous sections and discusses approaches that could increase integration of decision-support systems into development policy-making and natural resources management.

The contribution of decision-support systems to policy dialog depends on geospatial capacity (Box 7-3 and Chapter 8) (e.g., good data and equipment and trained staff) and communication among policy-makers, scientists, GIS experts, and civil society (Bassolé, et al., 2001). These interactions are "most effective within a fully supportive geo-information policy environment at the national level" (Bassolé et al., 2001).

In addition to geospatial capacity, demand will spur the development and use of decision-support systems. In the committee's opinion, the agricultural and natural resource management sectors are a likely primary source of this demand, as these sectors are the main users of geographic data and tools. The livelihoods of the majority of Africans depend on agriculture and natural resources, and pressing problems within these sectors include soil infertility and erosion, pollution from farm chemicals, pressure from grazing, and competition for resources. Addressing these problems demands better data and better ways of analyzing the relationship between human activities and changes on the land surface. Hence, decision support in the area of land cover (Chapter 6) will be one of the more fruitful application areas of geographic data and tools.

A further rationale for focusing on land cover is that it is basic information for many applications. Land use and cover change is at the nexus of a range of issues including habitat fragmentation, biodiversity, food and agriculture, water quality, urbanization and settlement, and human health and disease. Land cover is also readily obtained from satellites.

International activities could accelerate the use of decision-support systems for land-cover applications in Africa. For example, as the U.S. Geographic Information for Sustainable Development Alliance (GISD, 2002) initiatives are implemented, the need will arise to identify data for understanding land transformation processes. In addition, U.N. initiatives, notably the Global Land Cover Network and the GOFC/GOLD programs (Chapter 6), emphasize routine observations and analysis of land cover and change.

Strategies to improve or create these data sets are needed now, and these strategies should build on existing initiatives and networks. An effective land-cover decision-support system for Africa would include:

- **Development of standardized land cover and environmental classification systems.** Classification systems are central to the use of the product layers in a spatial data infrastructure. Classification systems are defined and formalized but are not rigid. FAO's Africover Land Cover Classification System (Chapter 6) is an emerging standard in Africa.
- **Development of a system for land-cover baseline and change detection across spatial scales.** This component would include a baseline map and compilations of change maps from repeated observations over time. These maps can be constructed using multiresolution remotely sensed data (e.g., AVHRR, MODIS, SPOT VEGETATION, ASTER [Annex Box 6-4], Landsat). Satellite observations alone cannot explain socio-economic and political factors that are among the causes of land-cover and environmental change, nor can they always identify trends or dynamics at the scale needed by decision-makers. Hence, remotely sensed data should be coupled with multi-scale geographically referenced economic and social data.
- **Identification of "hot spots" of change.** Because resources are scarce, and there are limited opportunities for decision-makers to make comprehensive evaluations, an approach that identifies areas of rapid change, high risk, or other critical occurrences should be developed. The routine identification of such hot spots will guide decision-makers to critical locations and times for the most efficient use of resources (e.g., Figure 6-5).
- **Analysis and modeling of the relationship of land-cover change to proximate causes.** Identification of

BOX 7-3
Enhancing the Contribution of GIS to Policy-making

Based on case studies in Burkina Faso, Côte d'Ivoire, and the Gambia, Bassolé et al. (2001) made the following recommendations:

1. Expand awareness of GIS value and use. Greater awareness is needed among African decision-makers of the value and usefulness of GIS analyses. This can be brought about through briefings and workshops where policy-makers have an opportunity to gain a better understanding of how GIS tools are developed and used.

2. Strengthen the policy dialogue process. The processes for encouraging and enhancing dialogue between policy-makers and GIS practitioners should be strengthened in all three countries studied. The process should be user-driven and should feature input from all levels of users (i.e., mid-level as well as senior government officials), and from civil society and other stakeholders.

3. Establish national geographic information policies. All African countries should establish national geographic information policies that address key issues and problems related to how this information is developed, accessed, and used. Some principal issues and problems include: how to enhance access to information and information sharing; setting the rules that determine standards and protocols for data collection, storage, labeling, and integration; data ownership; confidentiality and privacy; and copyright protection.

4. Strengthen local capacity. Capacity building is an essential component of developing national GIS strategies in West Africa. GIS capacity should receive more attention from the government and education sectors, and capacity should be developed and based in universities and the private sector as well as government agencies.

5. Expand access to geographic information. The results of GIS analyses—including databases, maps, and studies—should be made available to the public through the media, through regular government and private marketing and distribution channels, and increasingly through the Internet. Wider access will help strengthen the policy dialogue process by requiring policy makers to be more transparent in how they make decisions on resource pricing, allocations, concessions, revenues, and use.

6. Continue to study the use of GIS in policy-making. Each African country should conduct a periodic review of the development and use of GIS, taking into account the interests of the public, government, business, NGOs, academia, and other affected parties. In addition, more countries in Africa should develop case studies on the use of GIS in policy-making, thus building a stronger network of GIS users and experts.

drivers of change enables decision-makers to make critical interventions. Human causes of immediate, or proximate, land-cover change include land-management strategies that convert land cover from one type to another (e.g., from forest to field). A better understanding could be developed of the links between human activities and consequences for land-cover change if it were possible to measure the magnitude, frequency, and geographic distribution of proximate causes. With GIS and other analysis tools it is possible to develop quantitative analyses of spatial relationships associated with land-cover change and their drivers.

- **Definition of important pressures and remedies through case studies and modeling.** Direct observations of resource status alone do not reveal driving forces of land-cover and environmental change. If the drivers are unknown, natural resources management cannot be effective. Experience and knowledge for making policy decisions is gained by linking direct observations, case studies, and models to identify dominant drivers of environmental change. Linking observations at a range of spatial and temporal scales to empirical models that include socioeconomic activities and management goals allows a comprehensive, systematic approach to managing land-cover change.
- **Prediction of land-cover and environmental change risk zones.** Once a spatial model of environmental change has been calibrated, a projection can be generated representing the probability of future land-cover changes. This projection can be overlain on a land-cover map corresponding to the current situation to highlight areas at risk of unwanted change. Land-cover change probability maps allow validation of the models because they can also be used in retrospect to compare the actual land-cover changes that took place between two past observation years used in the model's calibration.
- **Development of environmental indicators.** An environmental indicator is a phenomenon or statistic associated with a particular environmental condition. Indicators of land use and cover change, and quality can be used to assess trends over time and to determine the efficacy of policy actions. Such indicators can be developed from various observation sources. Remotely sensed data can indicate areas where change is occurring. For example, the occurrence of fires or change in the spectral properties of data, which provides an indi-

rect "alarm" that there has been some kind of land-cover change). Subsequently, detailed analysis can be done through ground observations.

The agricultural and natural resource management sectors are among the main users of geographic information making land cover a priority area for the development of spatial decision-support systems in Africa. **A land-cover decision-support system should include a standard classification system; baseline data and change detection capabilities; hot spot detection and high risk zone prediction capabilities; analysis and modeling of proximate (mainly human) causes of change; linkages between direct observations, case studies, and models; and established environmental indicators.**

SUMMARY

Management of natural resources and development in a sustainable manner is ultimately a process of evaluation and decision-making. Decision-making is a complex process that involves value judgments and analysis of a broad array of information. GIS is a decision-support tool that can be used to integrate many kinds of data into a usable format, analyze data, and produce descriptive and predictive modeling of alternative scenarios.

Many decision-makers in developed and developing countries have no experience with GIS and other spatial decision-support tools, and thus do not appreciate their potential. Additional impediments to implementation of spatial decision-support systems include the orientation of projects toward data production rather than application; lack of planning for the decision-support process; lack of communication between technicians and scientists within an organization; and lack of inclusion of university research that could drive data analysis. Nonetheless, there are demonstrations of the value of decision-support tools in the African context such as the continent-wide MARA and FEWS NET programs, the Miombo Network in southern Africa, the Livestock Early Warning System in East Africa, and the Community Based Natural Resources Management program in Namibia.

Presently, there is a limited commercial market for geographic information, services, and technologies in Africa, but the need for spatial decision-support systems and demand is likely to grow. The agricultural and natural resource management sectors are among the main users of geographic information making land cover a priority area for the development of spatial decision-support systems in Africa.

Human, societal, and organizational capacity is needed to integrate geographic information and spatial decision-support tools into the decision-making process. The next chapter discusses geospatial capacity building.

REFERENCES

Bassolé, A., J. Brunner, and D. B. Tunstall. 2001. GIS: Supporting Environmental Planning and Management in West Africa. Washington, D.C. : World Resources Institute.

Chopak, C. 2000. Early Warning Primer: and Overview of Monitoring and Reporting. Available at: <http://www.fews.net>. Accessed June 28, 2002.

Cowen, D. J. 1988. GIS versus CAD versus DBMS: What are the differences? Photogrammetric Engineering and Remote Sensing 54:1551-1554.

Daratech. 2001. GIS software revenue tops $939 million in 2000. Press Release. Cambridge, Massachusetts: Daratech Inc. Oct. 24, 2001.

Densham P. J. and M. F. Goodchild. 1989. Spatial Decision Support Systems: A Research Agenda. Pp 707-716 in Proceedings GIS/LIS'89, Orlando, FL. Bethesda, MD: ASPRS

Eastman, J. R. 1999. Multi-criteria evaluation and GIS. Pp 493-502 in Geographical Information Systems, P. A. Longley, M. F. Goodchild, D. J. Maguire, and D.W. Rhind, eds. New York: John Wiley and Sons.

Eastman, J. R. 2001a. Guide to GIS and Image Processing. Volume 2. Worcester, MA: Clark Labs.

Eastman, J. R. 2001b. Uncertainty Management in GIS: Decision Support Tools for Effective Use of Spatial Data. Pp 379-390 in Spatial Uncertainty in Ecology: Implications for Remote Sensing and GIS Applications, C. Hunsaker, M.F. Goodchild, M. Friedl, and E. Case, eds. New York: Springer-Verlag.

EIS-Africa (Environmental Information Systems-Africa). 2001. Environmental Information Systems Development in Sub-Saharan Africa: Approaches, Lessons and Challenges. E. Gavin and J. Gyamfi-Aidoo, eds. EIS-Africa.

FEWS NET (Famine Early Warning System Network). 2002. Homepage. Available at <http://www.fews.net>. Accessed June 28, 2002

GISD (Geographic Information for Sustainable Development). 2002. Geographic Information for Sustainable Development homepage. Available at <www.opengis.org/gisd>. Accessed May 15, 2002.

LEWS (Livestock Early Warning System). 2001. Available at <http://cnrit.tamu.edu/lews/description.html>. Accessed May 20, 2002.

Malczewski, J. 1997. Spatial Decision Support Systems, NCGIA Core Curriculum in GIScience. Available at <www.ncgia.ucsb.edu/giscc/units/u127/u127.html>. Accessed June 5, 2002.

NCGIA (National Center for Geographic Information Analysis). 1990. Research Initiative 6: Spatial Decision Support Systems. NCGIA Technical Paper 90-5. University of California, Santa Barbara, September 1990.

NCGIA. 1996. Report from the specialist meeting on collaborative spatial decision making, Initiative 17. National Center for Geographic Information Analysis, University of California, Santa Barbara, September 1995.

NRC (National Research Council). 1999a. Our Common Journey: A Transition Toward Sustainability. Washington, DC: National Academy Press.

NRC. 1999b. Distributed Geolibraries. Washington, DC: National Academy Press.

Schmidheiny, S. with the Business Council for Sustainable Development. 1998. Changing Course: A Global Business Perspective on Development and the Environment. Cambridge, MA: MIT Press.

Simon, H. A. 1977. The New Science of Management Decision. Englewood Cliffs, N.J.: Prentice-Hall.

8

Building Capacity to Apply Geographic Information to Sustainable Development in Africa

INTRODUCTION

The effective use of geographic information to implement Agenda 21 will require sustained investments in human resources development, building public and private organizations, and improving societal capabilities for generating and using new knowledge. These efforts will involve strengthening existing international cooperation and introducing new approaches based on lessons learned from previous efforts. The application of geographic information to Agenda 21 issues in Africa requires considerable growth in geospatial capacity. Geospatial capacity is the ability to undertake activities, solve problems, and achieve objectives using geographic information and tools.

As countries transform themselves they have to develop different capacities. Capacity development needs to be addressed at three levels.

1. Human or individual—This involves enabling individuals to embark on a continuous process of learning—building on existing knowledge and skills, and extending these in new directions as fresh opportunities appear.
2. Organizational or institutional—This involves strengthening and improving existing institutions as well as the design of new ones where they do not exist. In additional to the role of individual organizations, emphasis is also placed on interactions between different organizations as a source of products and services derived from geographic information sciences.
3. Societal—This involves capacities in the society as a whole, or a transformation for development. An example is creating the kinds of opportunities, whether in the public or private sector, that enable people to use and expand their capacities to the fullest through reforms in policies, laws, and regulations. Societal capabilities are therefore linked to existing national systems of governance. Without such opportunities, people will find that their skills rapidly erode or become obsolete. If they find no opportunities locally, trained people will join the brain drain and take their skills overseas.

All of these layers of capacity are mutually interdependent and interactive; they co-evolve over long time horizons. For example, a paucity of organizations that employ a trained workforce may result in "brain drain" leading to the weakening of existing organizations such as universities, government departments, and private enterprises. When either human or organizational capacity-building is pursued on its own, development can become skewed and inefficient (Fukuda-Parr et al., 2002). Geospatial capacity-building is closely linked to expansion in economic activities and therefore part and parcel of the overall system of economic transformation. Policies to promote scientific and technological development will facilitate the use of geographic information for sustainable development.

This Chapter follows the above structural framework. (The committee chose to use the terms "organizational" and "human" rather than "institutional" and "individual.") The first section discusses human capacity-building, examining primary, secondary, post-secondary, and continuing or on-the-job training. Section two discusses organizations at levels from national to international (continental) including:

- governmental, civic, and private-sector institutions;
- academic and related institutions; and
- professional (scientific and technical) societies.

The final section treats the broader issue of societal geospatial capacity. It summarizes factors at the national level, including good governance,[1] that affect the applica-

[1]Governance is defined by the UNDP as "the exercise of political, economic, and administrative authority to manage a nation's affairs." Sound or good governance is defined as that sub-set of governance "wherein public resources and problems are managed efficiently and in response to the critical needs of society" (UNDP, 1997a).

tion of geographic information to sustainable development and discusses the role of partnerships in the development of geospatial capacity. The chapter discusses capacity-building research networks and organizations at the national, regional, and international levels. These organizations are described as examples and the committee does not analyze or critique their programs in terms of success or failure to build geospatial capacity.

HUMAN CAPACITY

The first manifestation of society's capacity is the competence of a critical mass of its citizens. Development of human capacity in a society is accomplished primarily through education and training. The basic health and nutrition of its citizens also determines a nation's human capacity (Box 8-1).

Developing geospatial capacity in Africa is part of the larger challenge of building scientific and technical capacity and a trained workforce. As in other scientific and technical fields, such as information and communications technology and agriculture (see Aiyepeku et al., 1994; Lindley et al., 1996; Cisse et al., 1998), geospatial capacity-building depends on

- primary education, including adequate nutrition and health care;
- secondary education, including interdisciplinary science and mathematics;
- post-secondary education and training; and
- continuing, and on-the-job training in relevant sectors.

Primary and Secondary Education

Capacity-building begins with primary education. By the secondary level multidisciplinary approaches may provide an avenue for learning about sustainable development. To be addressed effectively sustainable development issues require a multidisciplinary approach.

In sub-Saharan Africa enrollment rates in primary and secondary school are low and gender disparity is high. Currently girls are enrolled in lower proportion than boys in 26 sub-Saharan African countries (Sass and Ashford, 2002). African governments have made universal primary education by 2015 a major objective (NEPAD, 2001).

Opportunities for linking geographic science activities in secondary schools around the world are emerging. For example, a worldwide network of projects in secondary schools and universities called *My Community, Our Earth*[2] is helping students to use geographic information science to show how their communities are changing and how to make communities sustainable.

The integrated perspectives of geography and the technologies of the geographic information sciences are an integral part of the development and use of information and communications technologies in Africa and should be central to the new African Learning Network (ECA, 2001a). However, at present, geographic information is being developed primarily as an independent higher-level technical specialization in Africa. The same is true of the way geographic information science is being taught at the tertiary level throughout North America and Europe. Even where geographic information science is introduced as part of the secondary school system (such as in Ontario, Canada), the emphasis is on the technology itself rather than on how the technology can be used to deal with issues of societal significance.

Appropriate places in the secondary school curriculum include computer science and other information technology courses as well as interdisciplinary studies such as geography and environmental studies. The format of final secondary school exams is a key element in the integration of geographic information science in the curriculum.

Post-Secondary Education and Training

Post-secondary education and training[3] is particularly important for geographic information science and technology because of their scientific and technical natures. The importance of university education is stated in the NEPAD plan (NEPAD, 2001).

> The plan supports the immediate strengthening of the university system across Africa, including the creation of specialized universities where needed, building on available African teaching staff. The need to establish and strengthen institutes of technology is especially emphasized.

NEPAD's argument is supported by two studies that emphasize university education (Bourne, 2000; World Bank, 2000). NEPAD also calls for the establishment of "regional cooperation on product standards development and dissemination, and on geographic information systems" (NEPAD, 2001, p. 47). Such regional cooperation could promote capacity-building in geographic information science.

According to Bassolé (2002), 35-45 percent of human capacity in geographic information science takes place in the formal education system. Education and training in geographic information science in Africa is offered in universities or polytechnics in which geographic information science is part of a broader curriculum, and in national and regional training centers (Bassolé, 2002).

[2] <http://www.geography.com/sustainable/>.

[3] Post-secondary education and training refers to formal education at the tertiary level in universities, polytechnics, and in this case, regional remote-sensing centers.

BOX 8-1
Health Issues and the Workforce in Africa

Workforce issues cannot be discussed without reference to the immense impact of endemic disease on children and the working-age population (Gallup and Sachs, 1998). Good health is an integral part of human capacity-building. Diseases including HIV/AIDS, water- and vector-borne diseases (e.g., malaria, schistosomiasis), and tuberculosis have a profound impact on the workforce in sub-Saharan Africa. The effect of such losses among the working-age population is to exacerbate poverty and social disruption. Indeed, poverty and malaria are closely linked (Gallup and Sachs, 1998).

On the level of the individual these illnesses reduce energy and efficiency. On a larger scale diseases like AIDS are creating a generation of orphans and unraveling the fabric of society. The impact on the education system is devastating. When parents become ill or die, children are taken out of school, girls in greater number than boys, widening the gender gap in education. Two-thirds of the 113 million children out of school in developing countries are girls (World Bank, 2002). The problem is unlikely to abate: Although sub-Saharan Africa has two-thirds of the world's HIV infections and 84 percent of its AIDS deaths (UNAIDS, 2000), the region accounts for just 3 percent of global AIDS spending.

Although women now outnumber men infected with HIV/AIDS in sub-Saharan Africa, the biggest killer of women in Africa is not AIDS but AIDS-related tuberculosis. The United Nations warns that cases of tuberculosis in Africa are expected to double over the next 10 years. Such catastrophic losses to society undermine the potential for a stable workforce.

In West Africa 50-55 percent of training in the use of geographic information technologies occurs in the context of externally funded development projects that have a geographic information component (Bassolé, 2002). Often these projects work at the local level and demonstrate the potential of geographic information science in grassroots applications. On average they last three to five years until external funding expires. In addition, overseas development assistance often requires the use of technology, including hardware and software from the donor country. This results in the delivery of short-term intensive instruction on how to operate a particular software system but little knowledge of the concepts and principles of geographic information science. It may be challenging to transfer specific skills developed in project-based training to other situations. Human resource development does not always fit well within a project-oriented approach. Geospatial capacity-building is a long-term process.

Continuing Education

Geographic information technologies are evolving rapidly, therefore necessitating retraining. Kufoniyi (2001) says that "the rate of development in geographic information technologies is so rapid that it is often difficult for staff [of the Regional Centre for Training in Aerospace Surveys (RECTAS) in Nigeria] to be constantly retrained to keep pace with the rapid technological development."

Educational institutions in developing countries are beginning to address the needs for lifelong learning (NRC, 1999), and the Internet has a role to play. RECTAS, for example, is striving for full Internet connectivity to start a Web-based distance learning program that will include continuing education. *Our Common Journey* (NRC, 1999) underscores the role of the Internet in continuing education in developing countries. Internet-delivered courses in geographic information science are available from a variety of public and private sources. The use of the Internet and Web-based programs emphasizes the need for information and telecommunication infrastructure (ICT) plans (Chapter 4) involving the development of human capacity to manage this infrastructure. Internet delivery of education and training in geographic information science is an area of opportunity in Africa and for technical cooperation between African and donor nations.

The importance of ICTs for development was articulated by the Economic Commission for Africa (ECA, 1999) and recognized in the Okinawa Charter (Government of Japan, 2000). Two African programs assisting sustainable growth of human capacity in ICTs are the African Information Society Initiative (AISI) and the African Learning Network (ALN). The AISI provides a framework for information and communication activities in Africa, and the ALN is introducing ICTs to schools, linking universities and research organizations, and creating a national network for students enrolled in further education.

A cadre of well-trained individuals will need to be formed in each country to apply geographic data and information in support of sustainable development in Africa. **Continuing and on-the-job training should become an integral part of the process of enhancing geospatial capacity. Organizations that provide professional training in geographic information sciences such as regional centers and polytechnics should be strengthened.**

ORGANIZATIONAL CAPACITY

Within a country organizational capacity can be built on several levels—nationally, regionally, internationally—and simultaneously at multiple levels. The relative proportion of types of organizations differs among countries. For example, in developing countries there tend to be fewer formal organizations and more social and cultural groups (Fukuda-Parr et al., 2002). These social and cultural groups function similarly to formal organizations in developed countries, providing advice, assistance, and support. Because geographic information science is technical, geospatial capacity-building requires support from formal organizations in government, civil society, and industry. In Africa a number of programs to develop new types of organizations have been implemented to diversify capacity and promote coordination among various sectors (e.g., NEPAD, EIS-Africa, and SADC). One challenge facing organizations is the tendency for technically trained people to be attracted elsewhere (Box 8-2).

BOX 8-2
"Brain Drain" and "Brain Circulation"

"Brain drain" is an example of human capacity development resulting in a loss of organizational capacity. When staff are trained overseas, their withdrawal can weaken the organization if replacements are not found during their absence (Carrington and Detragiache, 1998; Eking, 1998; Downes, 2000). Where a development strategy takes brain drain into account, one beneficial approach might be to work with nationals in diaspora as a way of retaining links to the latest sources of scientific and technological advancement (Juma, 2000). For the adaptive and dynamic the world of the future could be characterized by "brain circulation," which would favor those countries that have placed a large number of nationals for training in centers of scientific and technological leadership. China is an excellent example of such strategic choices in capacity-building in geographic information science (Taylor, 1998). The situation in Africa, however, has been characterized by decades of institutional decay and low levels of enterprise development. These conditions have resulted in the absence of local absorptive capacity for expertise and added to the "brain drain". Greater private-sector investment will improve Africa's demand for expertise and help to reduce "brain drain". Appropriate organizational arrangements and incentives including research networks and regional cooperation could enable Africans in diaspora to contribute to geographic information activities in their home countries.

Role of Universities

In Africa, as elsewhere, universities and training centers play an important role in geographic education and training.[4] The most effective application of geographic information is carried out by individuals who understand both the technology and the socioeconomic development context in which it is to be applied (Akinyemi, 2001). Learning to apply modern geographic information and tools to address evolving societal needs requires a long learning period and attention to the development of research and analytical abilities as well as technical skills. Universities are the logical source of this kind of education and training because they focus on teaching and research. With the appropriate policies and incentives, universities are also natural incubators for enterprises and social organizations. The organizations of civil society are important in many African countries where the functions of the state are inadequate.

Training produces an educated workforce, and research is the basis for generating new knowledge and for developing products and services. Unfortunately, universities have tended to operate as discrete entities, focusing on teaching with limited interaction with wider society, including the private sector. Throughout the 1980s and 1990s massive funding cuts to universities reduced the research and teaching capacity of African universities (Ajayi et. al., 1996; Bourne, 2000; Association of African Universities and the World Bank, 1997; Davenport, 2000; Downes, 2000; Labatut, 2002; Mehta, 2000; Swartz, 2000; Task Force on Higher Education and Society, 2000). During this period, university enrollments increased but human and physical resources did not keep pace.

It is unlikely that long-term capacity-building in technical fields such as geographic information science can be sustained in the absence of strong foundations in higher education with emphasis on the science and technology. Despite the difficulties African universities face they remain vital to the generation of new knowledge and have the potential for organizational capacity-building. The application of geographic information to sustainable development in Africa will depend on the quality, character, and direction of university education in Africa.

There is an urgent need to coordinate and strengthen the capacity of university departments providing both research and training in geographic information science. **African universities should become a focus for capacity-building including training and research in geographic information science, and development organizations should coordinate their efforts to achieve this goal.**

National Organizations: Illustrative Cases

Bassolé (2002) stressed the importance of a supportive geographic information policy at the national level to encourage active partnerships among government, civil society, and industry. National-level capacity in developing countries has been particularly difficult to build (Fukuda-

[4]In Africa geographic information science is found in diverse faculties and departments, including geography, environmental sciences, and engineering. Applied research in geographic information science is often found in departments of geography, whereas research in remote-sensing, photogrammetry, surveying, geodesy, and other mapping sciences is usually found in faculties of engineering.

Parr et al., 2002), but examples exist. These include a program of integrated coastal management (Tanzania); an effort to use GIS in economic analysis (Uganda); and the mapping of freshwater resources (Burkina Faso).

Tanzanian Coastal Management Partnership

In 1997 the Tanzanian National Environmental Management Council (NEMC) formed a partnership with the University of Rhode Island and USAID to improve coastal management. Another goal was to strengthen the links between local and national coastal management agencies and the University of Dar es Salaam. Working groups were established in priority areas such as sustainable coastal aquaculture, tourism, and environmental monitoring, creating a bridge between coastal managers and the science community.

The partnership has resulted in several new programs. In 2002 the University of Dar es Salaam began offering five courses in coastal applications of geographic information. The partnership also produced a national coastal policy that is under review by the Tanzanian government. The Tanzanian coastal management partners have demonstrated organizational flexibility and commitment to apply geographic information and technology to coastal natural resource management.

Ugandan Information System

The National Environmental Information Center and Makerere University in Kampala[5] have developed GIS capabilities that can be applied to economic analysis and decision-making. The GIS project creates an application for a large volume of national social and environmental data that has been digitized at 1:50,000, with the goal of making this information available to the public. Makerere University formed the Advocates Coalition for Development and Environment (ACODE) to manage this GIS effort.

ACODE has developed an approach based on Principle 10 of the Rio Declaration. This says that states should encourage public awareness of environmental information, provide access to this information and to judicial remedy should access be denied. ACODE's objectives include the creation and support of an information access system for Uganda, common standards for information access across the country, and improvement of information exchange between industry and government.

A major challenge faced by ACODE in carrying out its work is inadequate access to government information. Access is limited by elaborate, time-consuming procedures for requesting information, poor government record-keeping that makes it difficult to track and locate information, and the hesitance of government officials who are unfamiliar with requests for access to information (Mwebaza, 2002).

Spatial Database of Water Resources in Burkina Faso

Burkina Faso is coordinating activities across internal boundaries and among government ministries and levels of government. A major challenge is the completion of a process that began in 1991 to coordinate the collection, production, and application of geographic information at the national level. Previously many applications of GIS were project driven and did not lead to a strategic plan or holistic vision for geospatial capacity-building. Currently Burkina Faso's goal with respect to geographic information is to establish a national policy that includes regulations for implementation, increase capacity for analysis in the universities, and raise awareness in the policy community.

A spatial database of water and natural resources in southwestern Burkina Faso was developed by the "Haute-Bassins" Regional Directorate of Hydraulics using more than 500 hydrological and natural resource maps of the region, large-scale remotely sensed Landsat images, and GIS to analyze and display data. The program includes 12 river basins and responds to Burkina Faso's need for basin-level management of water resources and for relevant nationwide government bodies that address water resource issues.

Demonstration of the wide array of GIS uses in governance has attracted the attention of other ministries within Burkina Faso and of neighboring countries, including Cote d'Ivoire, Mali, and Togo. In addition, a private sector GIS group, the Centre SIGET-A has been established to address natural resource management issues in Burkina Faso (USAID/WRI, 2001). The Centre SIGET-A emerged from the activities of EIS-Africa (Chapter 4). In turn, the Centre SIGET-A created a training program for students and is working to increase the demand for geographic information, tools, and services needed to invigorate the field of geographic information science in West Africa.

Regional Organizations

Regional organizations involve the cooperation of two or more countries in addressing common concerns such as shared resources. Organizations at the regional level permit countries to take advantage of economies of scale and provide opportunities for communication and collaboration. Regional cooperation promotes data standardization, and regional centers play a role in the continuing education of the workforce and raise awareness of policy-makers at all levels of the importance of geographic information.

The regional centers established by the ECA in 1964[6] are making an important contribution to increasing technologi-

[5]Specifically, the university's Makerere Institute of the Environment and Natural Resources.

[6]These centers were set up as a result of ECA Resolution 164 (VIII).

cal capacity in geographic information science. They include the Regional Centre for Mapping of Resources for Development (RCMRD) in Kenya (Box 8-3), RECTAS in Nigeria, the Southern and Eastern Africa Mineral Centre (SEAMIC) in Tanzania, and the African Centre for Meteorological Applications and Development (ACMAD) and Agro-Hydro-Meteorological Center (AGRHYMET) in Niger.

At RECTAS there are four program components: two master's level courses, a three-month certificate course, and refresher courses. RECTAS also provides consulting services and short-term customized training. Until 2000 RECTAS's main contribution to capacity-building was to develop a technical workforce in photogrammetry and remote-sensing. Recently the center reoriented its activities to contribute to the development of both human and societal capabilities in the region. As the director of RECTAS observed, "The need for capacity-building in geographic information production and management in Africa cannot be over-emphasized as geographic information is definitely the sine-qua-non for sustainable national development" (Kufoniyi, 2001).

Continent-Wide Organizations

Several organizations are approaching capacity-building on a continental scale (e.g., EIS-Africa, the Africa Environment Outlook [AEO] Project and ECA). Continent-wide efforts draw organizations from multiple sectors together and promote open access to data and information.

EIS-Africa is applying GIS and remote-sensing to natural resource management and development issues in Africa (Chapter 4). As a network for the cooperative management of environmental information, EIS-Africa draws together private- and public-sector organizations and experts, promoting access to and use of environmental information (Bassolé, 2002).

In 2000, the African Ministerial Conference on the Environment approved the production of the AEO report involving participation of UNEP and multiple partners in six regions (central, eastern, northern, southern, and western Africa and the western Indian Ocean islands). This project harmonizes regional and national-level sources of environmental data and is striving to build capacity in the area of environmental reporting, policy analysis, and scenario development. For example, GIS has been used to identify high-priority biological conservation areas across the continent (AMCEN, 2002).

Recently ECA reorganized its structure and processes to promote effective interaction among various groups interested in information for development. ECA's Committee on Development Information (CODI-2) developed a plan for capacity-building entitled "The Future Orientation of Geographic Information Activities in Africa" (ECA, 2001b). This document identified as a key problem the lack of individuals trained in geographic and interdisciplinary science. Drawing together previously separate activities, ECA has created a new organizational structure with increased potential for capacity-building.

Research Networks

Research networks represent an important avenue for using existing capacity, focusing their goals on meeting local and international needs, for promoting "brain circulation"

BOX 8-3
Regional Centre for Mapping of Resources for Development, Nairobi

Focus: The RCMRD trains leaders in planning and decision-making, project implementers, trainers, and researchers. Its remote-sensing and environmental management program addresses environmental problems through training in a number of activities, including maintenance of a regional early warning system for food security, environmental monitoring, and disaster management. RCMRD's engineering services program is expanding the center's capacity and capability to service and maintain automated equipment and hardware used in resource-mapping and environmental management. The focus of RCMRD has evolved to reflect changing technology, changing geospatial capacity in the region that it serves, and changing socio-economic and operational needs of the member countries (Figure 8-1).

Sponsor/partners: The RCMRD operates with funds from contracting member states and from donors, which include USAID, UNDP, FAO, World Bank, IDRC, UNESCO, BADEA, and a number of bilateral donors, such as France, India, Italy, and the Netherlands.

Key results to date: The center has trained more than 3,000 people from member states and other African countries. Former trainees now train others and provide technical services in their own countries.

Lessons learned: Confronted with a decline in funding from member states, the center was advised in 1997 by UNEP, UNDP, and the World Bank to reorient its efforts to problem-solving applications in natural resource development and environmental management. Commercialization of services was recommended as a means of ensuring long-term financial stability of the center, and the center was encouraged to partner with networks such as EIS-Africa to promote sharing of knowledge and best practices.

SOURCE: W. Ottichilo (RCMRD, personal communication, 2002)

FIGURE 8-1 Contracting and non-contracting member states of the Regional Centre for Mapping of Resources for Development in Nairobi (courtesy of RCMRD).

(see Box 8-2), and providing opportunities for young researchers to enhance their capabilities (e.g., Miombo Network, Chapter 7). These networks build upon ongoing research activities in universities, government, industry, and civil society organizations. Donor-funded networks often are created to promote the effective use of available resources and succeed where demand exists for their services. Networks do not produce demand; rather their effectiveness depends on the degree to which demand is incorporated into their planning.

International Organizations

Although indigenous geospatial capacity-building efforts are growing, international players dominate the application of geographic information science to development in Africa. Fukuda-Parr et al. (2002) describe how traditional capacity-building initiatives adversely affected the development of local organizations for training and education because they tended to

(1) undermine local capacity,
(2) distort priorities
(3) focus on high-profile activities,
(4) fragment management,
(5) use expensive methods,
(6) ignore local wishes, and
(7) fixate on targets.

These problems of technical cooperation were a result of two mistaken assumptions in particular according to Fukuda-Parr et al. (2002):

> The first is that it is possible simply to ignore existing ca-

> pacities in developing countries and replace them with knowledge and systems elsewhere—a form of development as displacement rather than development as transformation. The second assumption concerns the asymmetric donor-recipient relationship—the belief that it is possible ultimately for donors to control the process and yet consider the recipients to be equal partners.

International agencies that build geospatial capacity in Africa have learned from these lessons and are paying greater attention to the central role of local capacity in development. Although the influence of donors will continue to determine the direction of capability-building efforts, there is considerable scope for taking into account recipient needs without compromising the requirements for accountability among donor agencies. For example, extending the timeframes for donor projects may have a greater impact on capacity development than simply increasing the level of funding available.

Examples of international organizations that are active in African geospatial capacity-building are USAID; the International Development and Research Center (IDRC), Canada; the Aerospace Remote Sensing Development Group, France (GDTA); and the International Institute for Aerospace Survey and Earth Sciences (ITC), the Netherlands. Many of the examples of USAID's work are introduced in Chapter 3 and subsequent Chapters. The IDRC supports development research and contributes to building research capacity in African universities in ICT. Geographic information processing is an integral part of this effort through its Acacia program (IDRC, 1999; Labatut, 2002). GDTA and ITC are partners of RECTAS in Nigeria. Established in 1973 as an economic consortium, GDTA operates mainly in Francophone Africa. It trains people in remote-sensing and GIS through courses and workshops in France or in the country requesting the training. In 1950, the Netherlands Government founded ITC (Box 8-4) at the request of the United Nations to build capacity through educating and training mid-career professionals from developing countries. Initially the institute concentrated on photogrammetry and cartography but as technology in the mapping sciences developed new activities were added, including training in the analysis of satellite imagery and GIS.

Among other international groups contributing to capacity-building in Africa are universities and colleges and professional organizations. In recent years international professional associations in the geographic information sciences such as the African Organization for Cartography and Remote Sensing, Federation Internationale Géographique, EIS-Africa (Chapter 4), International Society for Photogrammetry and Remote Sensing, International Cartographic Association, International Association of Geodesy, and International Hydrological Organization have held meetings allowing local professionals access to the latest developments in the field. Universities and colleges in Europe and North America provide scholarships to Africans to study geographic information science (e.g., the Fulbright scholarship program). However, these organizations may also be detrimental to capacity development because they reduce the incentive for donors to support the creation of similar organizations in Africa. Indeed, the large number of international organizations operating in Africa may reduce the potential for the emergence or maturation of local organizations. International organizations often operate under an immune policy environment and, therefore, have no incentive to advocate public-policy reforms that promote the wider use of geographic information. An appropriate balance needs to be struck between the role of international organizations and the need to create space for the emergence of local public and private organizations.

SOCIETAL CAPACITY

Cooperation among all sectors of society (e.g., government, civil society, and the private sector) is essential for the development of geospatial capacity. Government's role is central. U.N. Secretary-General Kofi Annan said,

> All our work for development and peace has taught us that if the issue of governance is neglected, then we are building on sand. No amount of aid, no degree of diplomacy can produce lasting progress if it is not rooted in legitimate, rule-bound institutions responsive and accountable to the people (Annan, 2002).

Government has the power to formulate policies that encourage access to data and information, facilitate relationships among the three sectors, and create an environment in which private-sector development can flourish. In turn, the private sector builds demand for geographic goods and services. Although demand currently exceeds supply in geographic information science in Africa, overall demand for geographic information, goods, and services is very low. The critical mass of individuals required for the development of societal capacity in geographic information science will not enter the field in the absence of demand that creates jobs and income.

This section discusses the role of science and geographic information in governance, the societal factors that influence geospatial capacity, and partnerships for geospatial capacity-building within Africa and between African countries and the United States.

Science and Governance

A government's recognition of the value of geographic information for policy-making reflects the level of attention given to scientific and technological issues in general. Since the adoption of Agenda 21 in 1992, emphasis on scientific and technical capacity-building has increased in Africa. This increase likely is related to the following:

BOX 8-4
The International Institute for Aerospace Survey and Earth Sciences

Since the 1960s, ITC has been in the forefront of curriculum development for geographic information sciences in Africa. Recently the organization reframed its curriculum and program goals to address Africa-centered issues. Students can receive master's or Ph.D. degrees or can attend short refresher courses. Between 1950 and the present, ITC graduated more than 4,000 students from 45 African countries. The largest number of graduates were from Tanzania (553) followed by Ethiopia (499), Nigeria (442), Kenya (382), Sudan (309), and Egypt (308) (Beerens, 2002). ITC's decentralization strategy illustrates a number of important challenges shared by all aid agencies. These include:

- mobilizing adequate resources for regional centers and universities,
- creating organizational linkages within countries and regions, and
- promoting geospatial capacity-building without competing with indigenous efforts.

Competing with indigenous efforts is contrary to ITC's goals to build geospatial capacity in Africa, yet "it is easier for some Nigerians to study cartography in The Netherlands than at Kaduna Polytechnic" (Ademlemo et al., 1985). According to Beerens (2002),

> Perhaps donors themselves should not set standards too high, standards that require African countries to look for help from the outside, either in the form of expatriate technical assistance or overseas education and training. The problem then is that this type of assistance, although at first temporary and targeted, becomes structural. We have to accept that development takes place not by throwing money, projects and expatriate technical assistance at problems but by recognizing the need to start from local conditions and capacities.

- Responsibilities of African countries as signatories of international treaties for reporting in areas of science ranging from agriculture and natural resources to climate change (Cisse et al., 1998).
- Recognition by African and donor nations that capacity-building and good governance are necessary for economic reform (UNDP, 1997a,b; Nsouli, 2000).
- Explosion of the Internet that brought concerns about the "digital divide" to the forefront.

Because Africa is a large and diverse continent, there are significant national differences in approach to the application of geographic information. The capacity to produce and use geographic data varies between and within countries. For example, South Africa has a sophisticated research and educational system for producing geographic data expertise and a long tradition of using geographic information in administrative and policy contexts. In other African countries these efforts are just beginning. African countries that do not have the technical expertise and infrastructure to gather and process geographic data rely on access to geographic information from other countries.

Mohammed Hassan, executive director of the Third World Academy of Sciences and president of the African Academy of Sciences, made these remarks about the need for science advice for African governments.

> Africa, a continent with nearly 1 billion people, has less than 30,000 African-born Ph.D. scientists living and working there...It is clear that both well-trained scientists and strong scientific institutions are in short supply in Africa and that the absence of one helps to explain the absence of the other...More funds must be invested in academic research and training activities and academies must play a larger, more authoritative role in advising their governments (Hassan, 2001).

In African governments recognition is growing of the need for science capacity in a world defined by economic globalization and information (NEPAD, 2001; Rabenoro, 2001). Evidence is needed at all levels of government that science capacity can contribute to economic competitiveness and human well-being. To promote the use of geographic information for sustainable development governments need coherent science, technology, and innovation policies with the appropriate institutional arrangements for science and technology advice to key branches of government. Efforts to educate policy-makers about the potential applications of geographic information and technology are underway in many countries including the United States.[7] Practical applications in areas of transportation, cadastral issues, and disaster mitigation are among the examples used to illustrate the contributions of GIS to the business of government.

Geographic Information and Governance

The relationship of geospatial capacity and good governance in Africa hinges on broad public access to information and the decision process and accountability on the part of the government regarding decisions including the allocation of public resources such as food, land, and water. Resource allocation is central to African development efforts that focus on the eradication of poverty (e.g., NEPAD, 2001). Geographic data provides governments with needed information

[7]Tap into the Power of GIS. Available at< http://www.fgdc.gov/nsdi/docs/communications>.

about territory, spatial patterns (e.g. population distribution or urban forms), flows of people and goods, and human-environment relations (Murphy, 1995).

Civil society plays an important role in Africa. Non-governmental organizations assume responsibility for providing food, clothing, and health services in many countries (Joseph, 2002). Organizations such as ACODE in Uganda (Chapter 8) provide information to the public and are part of a system of checks and balances for the government. Traditional organizations such as the *gachacha* in Rwanda and *kgotla* in Botswana provide a local arena for debate and dispute resolution helping to lighten the burden on the formal legal system (Cliffe, 2002).

Geographic information and good governance go hand in hand. Access to integrated (social, environmental, and economic) geographic information allows civil society to hold government accountable for its decisions; government creates policies that allocate goods and services, and determine public access to information and public participation in the decision process.

The concept of spatial data infrastructures has existed for about a decade (NRC, 1993). Its value to society has not been fully realized (Chapter 4). Focus groups and courses to acquaint national policy-makers with spatial data infrastructures were among the suggestions offered at a meeting of the GISD Alliance[8] in Nairobi in 2002 (E. Gavin, personal communication, 2002) and echoed by Bassolé (2002).

National governments are key players in applying geographic information to sustainable development. One of the critical decisions that African governments make relates to how well or fairly public resources are managed and shared. Governments also have the power to enact legal and regulatory frameworks to ensure availability of geographic data for use in government operations and in policy- and decision-making. International reporting responsibilities require African governments to establish frameworks from the grassroots to the national level to inventory and monitor the state of the country and its resources. Geographic information can help African governments to meet their national and international reporting responsibilities incurred as signatories of treaties such as the Convention on Biological Diversity and the Convention on International Trade in Endangered Species.

Governments can provide incentives for the use of geographic information science at the sub-national level through legislation designed to achieve sustainable development goals. In Namibia the Ministry of Environment and Tourism's CBNRM (Chapters 3 and 7) promotes sustainable use of natural resources and facilitates communication between the national government and rural communities on wildlife and tourism issues using two-dimensional maps and images. Currently there are 14 conservancies in the program and 14 more are interested in joining. The devolution of environmental monitoring responsibilities by the national government to provincial or local governments promotes the demand for and use of geographic data at sub-national levels. While opportunities for open access to geographic information are increasing with efforts towards democratization and with advances in the Internet that make digital maps widespread, programs that engage communities and local citizens and land managers with paper maps and other accessible outlets for these geographic information can promote broader participation in the decision process at the local level.

[8]This meeting included participants from Algeria, Botswana, Ethiopia, Kenya, Lesotho, Malawi, Morocco, Mozambique, Namibia, Seychelles, South Africa, Uganda, Zambia, and Zimbabwe.

Social Factors Influencing Geospatial Capacity

Data, hardware, and software systems provide increasingly sophisticated geographic information worldwide, but it is really the political, social, economic, and educational institutions of a country that ultimately determine the application and use of geographic data for decision-making. Nonetheless, this barrier could be overcome through building social capacity. Awareness of the importance of geographic information to Agenda 21 issues is a basis for developing social capacity; so too are the efficient management of geographic information and access to information and inclusion in the decision process.

The Efficient Management of Geographic Information

Spatial data infrastructures whose standards conform to the Global Spatial Data Infrastructure (GSDI) have the potential to enhance the use of geographic information by society. With shared compatible data, countries can monitor and manage areas that transcend national boundaries such as river basins and forest systems. Seamless geographic information can facilitate discussions about shared resources. Participants at the Geographic Information for Sustainable Development Alliance meeting in Nairobi (E. Gavin, personal communication, 2002) identified major difficulties in managing geographic information: lack of a clear policy framework, lack of standardization, and technical constraints (see also NSIF, 2000). Addressing these challenges requires collaboration among all sectors and acceptance of the principles underlying spatial data infrastructures (Chapter 4).

Access to Information and Inclusion in the Decision Process

As the goal of universal education in all countries is pursued a growing number of people will be able to use information to inform themselves about government activities. People want access to information about their environment, their health, and their economic and social opportunities. Information provides people with more control over their lives and the quality of their environments. Access to information in a country includes the right to information and the right to participation in decision-making (Mwebaza, 2002). Ideally, if access to data were denied, legal processes to redress grievances would be in place.

The rapid development of new information technologies and networks over the past two decades has made it possible to obtain geographic data through remote-sensing and to store them in databases for later use. Detailed data about a country can be obtained by other countries and by commercial firms without the observed country's permission. Databases can be created wholly outside the country of interest and the information could be of superior quality to that available in the country of interest. This growing capacity of individuals and organizations to obtain data from global network sources means that governments can no longer completely control access to data and information about their countries. As awareness of information grows access to information will be in greater demand.

Decision-making involves both objective information and subjective goals and values. The goal of a spatial decision-support system is to improve decision-making through an informed process, not to encapsulate all facets of the process. Although geographic information can improve decision-making, adding information to the process without increasing public access can exacerbate the growing digital divide.

In an increasingly technical world where decision-making can be based on vast and complex databases the ordinary citizen can fall behind the "information power curve." This is particularly true for the application of geographic data that may require powerful computers and trained technicians. Often projects and programs using GIS technology have converted paper maps to computer maps that are subsequently stored in central computer databanks, often inaccessible to all but the technically trained agency personnel. With the emergence of digital processing of geographic information, there is a risk of further increasing the digital divide as new analyses and models are implemented without wide public access to the inputs and results.

Partnerships for Geospatial Capacity-building

Geospatial capacity-building, like the transfer of technology discussed in Chapter 7, will be more effective when the cooperation occurs in the context of long-term, practical partnerships such as business-to-business or university-to-university rather than exclusively in the traditional donor-agency-to-recipient-government partnership. Partnerships among those with similar experiences (e.g., between entities in developing countries) could enhance their efficacy. Over time, a commitment is needed for communication and education of partners and for human and organizational adjustments to the use of new technologies (Schmidheiny, 1998).

In addition to partnerships driven by demand for specific services, entrepreneurship defined broadly to include social, business, and community activities, is a driving force behind the development of capacity for using geographic information in the sustainability transition. This section looks toward a more sustainable and dynamic approach to capacity-building that creates an important role for African universities.

New Partnerships

Effective use of geographic information science in sustainable development will be associated with the emergence of partnerships involving universities, industry, government, and civil society.

University Partnerships

Universities bring together many of the elements needed for innovative partnerships. The growing need to build human and organizational capacity to address issues of environment and development offers universities new opportunities to expand their teaching and research. Universities can play a role in offering conservation education and in helping to create conservation organizations. The Stellenbosch University in South Africa is promoting the use of geographic information science in Africa with its satellite launched in 1999 in cooperation with NASA. This is the world's only satellite built and managed by a university (Box 8-5).

Focusing on capacity-building through research universities may entail changes in the way African universities function. African universities can contribute to societal capacity by functioning as nodes in global networks of knowledge. Universities can provide the knowledge and the know-how that government and the private sector need. Government and the private sector, in turn, can create social and practical support for universities.

A dual focus on teaching and research would build and retain capacity, providing the incentive for African scholars to remain in African universities. These changes should be promoted from within universities, and not imposed on them. A new generation of partnerships among government, civil society, and the private sector could foster innovation and entrepreneurship in Africa.

Entrepreneurship: Developing Demand

In Africa, supply and demand in geographic information science are often out of balance. A study of the situation in Nigeria in the mid-1980s (Adeniyi, 1985; Ademlemo et. al., 1985; Duru, 1985) shows that the shortage of trained personnel limited the country's capacity to effectively use them. Of 93 Nigerians trained externally in remote-sensing (many at ITC) only 22 were in a position to apply remote-sensing techniques and only five of them had the necessary equipment for their work (Adeniyi, 1985). Partnerships of universities and the private sector in geospatial capacity-building are key to achieving a balance between supply and demand for geographic information, tools, and services in Africa.

Countries are searching for ways to facilitate the transformation of new knowledge into products and services, and

BOX 8-5
Universities as Incubators:
The Case of theUniversity of Stellenbosch

The Stellenbosch University Satellite (SUNSAT) program set out in 1992 to enrich the engineering graduate training program; expand international scientific cooperation; and stimulate interest in technical careers among school children.

The venture started in the Computer and Control Systems Group of the Electrical and Electronic Engineering Department of the university to set up a post-graduate research group in satellite systems. This resulted in the establishment of the Electronic Systems Laboratory (ESL) in the department in 1992. SUNSAT was designed, constructed, and tested entirely by ESL students and staff, with the exception of the solar arrays, batteries, GPS receiver, and laser reflectors.

The Department of Communication of the South African government has identified the need for more engineers to supply the demand for satellite telecommunication services and the University of Stellenbosch is meeting this challenge. The impact of SUNSAT on capacity-building can be measured by the more than 50 master's and Ph.D. degrees that have been awarded to students who participated in the satellite's development.

SOURCE: <http://www.sunspace.co.za/_index.html>.

closer cooperation between universities and the private sector has emerged as one of the most efficient ways of achieving this goal. Governments can also collaborate with universities to provide education for managers and policy-makers. Private-sector participation in geographic information science will require the support of government for a wide range of enabling policies and incentive measures. Universities could play a key role by providing much needed research and development to start-up geographic information enterprises in both civil society and in the private sector.

Private-sector demand in Africa as a whole is likely to grow fastest in the areas where the present users of geographic information and tools are concentrated: agriculture and natural resource management (Chapter 7). Data and tools for agricultural and environmental issues are needed to address local concerns, including poverty, land ownership, water resource management, and the relationships between competing stakeholders. Capacity-building is needed in both the formal and informal agricultural sectors; extension workers need training that emphasizes both skills and knowledge (Lindley et al., 1996). Agriculture and natural resource management are areas of opportunity for entrepreneurship involving geographic information science and technology.

Partnerships between universities and industry will not emerge automatically. Efforts to create novel partnerships without government incentives and support are unlikely to meet with success. Strategic alliances among government, civil society, and the private sector are a global phenomenon and represent a significant source of productivity and dynamism. The prospects for such alliances in the fields of geographic information are immense and need to be tapped.

In promoting organizational cooperation emphasis should be placed on fostering innovation and the transfer of geographic data and technology through: (1) partnerships and research networks among government agencies, research and training institutions, the private sector, and the non-governmental sector; (2) international collaboration involving developed and African countries; and (3) cooperation between African and other developing countries.

U.S.-Africa Partnerships

Growing recognition of the role of science and technology in development is creating opportunities to redefine international diplomacy. Good diplomacy entails deploying the benefits of science and technology to meet the needs of developing countries. The United States is already engaged in providing support to African countries for the development of their spatial data infrastructures (see Appendix C). At the U.S. National Academy of Sciences meeting on April 30, 2002, U.S. Secretary of State Colin Powell made the following remarks about the importance of science and technology to sustainable development.

> You will also see our new approach to development at the World Summit on Sustainable Development in Johannesburg, South Africa, this August and early September. At the summit, we will stress that good governance, including solid science and technology policies, are fundamental to sustainable development. We will also emphasize in Johannesburg that as important as government-to-government cooperation is to development, governments alone cannot do the job. Public-private partnerships will be crucial to find the money needed to help nations address the daunting problems that they face in developing.

The United States and many other countries and multilateral organizations are working to improve the geospatial capacity of African countries. For example, the United States is engaged with African countries in organizational partnerships promoting the use of geographic information including the Open GIS Consortium Pilot Project (Box 4-4) and the nine USAID Collaborative Research Support Programs [Table 3-1]). The success of these efforts depends on whether they (1) respect the national basis both for African government operations and decision-making and for local non-governmental organizations; (2) provide for open access to data and information and the use of common standards to promote data integration; (3) provide reasonable guarantees of stability and continuity in financial support; and (4) recognize the importance of traveling the "last mile" from the technical data to the non-technical decision-maker.

In addition to the many benefits that African countries derive from access to and use of geographic data, the United States benefits from improved capacity in Africa to use geographic data and information. Improvements in access, management, and use of geographic data in Africa can promote global stability through participatory government and help African countries to fulfill their international obligations incurred as signatories of international treaties. Information management in African countries also improves the data available to scientists and policy-makers in the United States for scientific research, economic activities (such as the expansion of consumer markets in Africa for U.S. goods and services), and for the care of the global commons: the oceans and the atmosphere.

In addition to supporting developing countries in the development of their spatial data infrastructures, the United States provides valuable foundation and thematic geographic data for a wide variety of applications in Africa. These data include

- free and open access to the 24-satellite Global Positioning System (GPS) (Chapter 5);
- global 30 × 30 m orthorectified Landsat Thematic imagery from circa 1990 through the NASA Data Buy and Earth Satellite Corporation (Chapter 6);
- imagery of many African countries from CORONA data and space shuttle photography;
- global digital elevation model information (at 90 × 90 m spatial resolution) derived from NASA's SRTM radar data (Chapter 5);
- hydrologic information derived from GTOPO30 (Chapter 5);
- land cover derived from Landsat, Terra MODIS, and AVHRR imagery (Chapter 6);
- remote-sensing-derived vegetation indexes, including the NDVI used in the Famine Early Warning System and elsewhere (Chapter 6);
- tropical rainfall measurement from the Tropical Rainfall Measuring Mission (Chapter 6);
- soil moisture measurements from the Defense Meteorological Satellite Program;
- estimation of human population distribution using LandScan 2000 and Gridded Population of the World datasets (Chapter 5); and
- fire monitoring using DMSP nighttime city lights and Terra MODIS imagery (Chapter 6).

As a player in the development of geospatial capacity in Africa the United States could more effectively build on existing activities in ways that are compatible with African needs. These include building capacity for long-term use of geographic data, bridging the gap between technical and policy/operational users, supporting and promoting long-term partnerships between African and U.S. universities,[9] and recognizing the role of national and local organizations in Africa in developing demand for geographic information services and applications.

[9]For example, USAID's Collaborative Research Support Programs (CRSPs), (Chapter 3).

SUMMARY

The application of geographic information to Agenda 21 issues in Africa requires considerable growth in geospatial capacity. Capacity development is addressed at three interdependent levels: human, organizational, and societal. Human capacity is accomplished primarily through education and training. Owing to the technical nature of geographic information science, university and on-the-job training are vital to enhance geospatial capacity. Among the challenges faced by organizations in building capacity are retention of trained people and restricted access to information. An appropriate balance needs to be struck between the role of international organizations and the need to create space for the emergence of local public and private organizations. Societal capacity benefits from partnerships among universities, governments, and the private sector, and from open access to data.

The final chapter summarizes lessons learned in the application of geographic information in Africa and offers the committee's conclusions and recommendations.

REFERENCES

Adeniyi, Peter, O. 1985. Remote Sensing, Resource Development and Education in Africa. Pp. 189-224 in Education and Training in Contemporary Cartography. D. R. F. Taylor, ed. Chichester, U.K: Wiley.

Ademlemo, I. A., O. O. Ayeni, and D. Y. Balogun. 1985. Manpower and Curriculum Development in Cartography in Developing Nations. Pp. 227-242 in Education and Training in Contemporary Cartography. D. R. F. Taylor, ed. Chichester, U.K: Wiley.

Aiyepeku, W.Q., A. Iwayemi, and I. Ajiferuke. 1994. Information Technology and Capacity Building for the Global Environment (Africa). ARCIS consultancy report. Global Environment Facility (GEF) October 1994. Available at <http://www.sas.upenn.edu/Africa_Studies/Padis/AiyebExe.html>. Accessed May 19, 2002.

Ajayi, J. F. A., L. K. H Goma, and G. A. Johnson. 1996. The African Experience with Higher Education. Ghana: Association of African Universities.

Akinyemi, F.O. 2001. Geographic Targeting for Poverty Alleviation in Nigeria: A Geographic System (GIS) Approach. Proceedings of the 20th International Cartographic Conference. Vol. 2. Beijing, China.

AMCEN (African Ministerial Conference on the Environment). 2002. The African Environment Outlook: Past, Present, and Future Perspectives. The AEO project. Available at: <http://www.grida.no/aeo/005.htm>. Accessed July 8, 2002.

Annan, K. 2002. The Spread of Democracy Offers Africa's Best Hope for Recovery. Speech delivered at Harvard University, Kennedy School of Government 25 April 2002. Available at: http://www.ksg.harvard.edu/news/annan_forum_042502.htm. Accessed May 20, 2002.

Association of African Universities and the World Bank. 1997. Revitalizing Universities in Africa: Strategies and Guidelines. Washington, DC: World Bank.

Bassolé, J. 2002. Presentation at National Research Council, Washington, DC, January 10-11, (a) West Africa: Workforce Issues in the Context of Capacity Building and Decision Making; (b) Upper Niger Region: Contribution of GIS and RS to Addressing NRM and Development Issues. Available at <http://www.opengis.org/gisd/cgi/listgisdtxt.pl>. Accessed June 5 2002.

Beerens, S. J. J. 2002. Capacity Building for Geospatial Information Handling in Africa: The ITC Perspective. Presentation to Committee on the

Geographic Foundation for Agenda 21. Washington, D.C., January 2002.

Bourne, R. (ed). 2000. Universities and Development. London: Association of Commonwealth Universities.

Carrington, W. J. and E. Detragiache. 1998. How Big is the Brain Drain? IMF Working Paper 98/102.Washington, DC.: IMF.

Cisse, M.K., Y. Sokona, and J.-P. Thomas. 1998. Capacity Building: Lessons from Sub-Saharan Africa. Paper presented at the International Conference on the Implementation of the United Nations Framework Convention on Climate Change in Africa. Accra, Ghana, 8-10 July 1998. Available at <http://www.edna.sn/energie/acra-eng.htm>. Accessed May 19 2002.

Cliffe, L. 2002. "African Renaissance?" In Africa in Crisis: New Challenges and Possibilities, T. Zack-Williams, ed. London and Sterling, Virginia: Pluto Press.

Davenport, P. 2000. Development and the Knowledge Economy. In Universities and Development, E. Bourne, ed. London: Association of Commonwealth Universities.

Downes, A. 2000. University Graduates and Development. In Universities and Development, E. Bourne, ed. London: Association of Commonwealth Universities.

Duru, R. C. B. 1985. Status and Constraints of Automated Cartography Training in Africa: The Nigerian Example. In Education and Training in Contemporary Cartography. D. R. F. Taylor, ed. Chichester, U.K: Wiley.

ECA (U.N. Economic Commission for Africa). 1999. Strengthening Africa's Information Infrastructure. Report from the African Development Forum. Available at <http://www.uneca.org/adf99/infrastructure.htm>. Accessed January 28, 2002.

ECA. 2001a. The African Learning Network: Emerging from Behind the Knowledge Curtain. E/ECA/DISD/CODI.2/21. Addis Ababa. UNECA.

ECA. 2001b. The Future Orientation of Geoinformation Activities in Africa: Synthesis. Available at <http://www.uneca.org/eca_resources/Conference_Reports_and_Other_Documents/disd/codi/docs/doc3EN.pdf>. Accessed May 15, 2002.

Eking, D. 1998. Sustainable Development and Graduate Employment: The African Context. In Graduate Prospects in a Changing Society, A. Holden Ronning and M. Kearney, eds. Paris: UNESCO Publishing.

Fukuda-Parr, S., C. Lopes, and K. Malik. 2002. Capacity for Development: New Solutions to Old Problems. London: Earthscan Publications.

Gallup, J. L. and J. Sachs. 1998. The Economic Burden of Malaria. Center for International Development, Harvard University. WHO/28 Press Release. April 24, 2000. Avalable at <http://www.mofa.go.jp/policy/ecnomy/summit/index.htm.>. Accessed August 29, 2002.

Government of Japan. 2000. The Okinawa Charter on the Global Information Society, Kyushu-Okinawa G8 meetings.

Hassan, M. 2001. Remarks at the Interacademy Panel. Capacity Building for Academies in Africa. Trieste, Italy. May 16-18, 2001.

IDRC (International Development Research Centre). 1999. Internet for All: The Promise of Telecentres in Africa. IDRC Briefing. Paper 3, Ottawa, Canada.

Joseph, R. 2002. Smart Partnerships for African Development: A new strategic framework. Washington, DC: United States Institute of Peace. Available at <http://www.usip.org>. Accessed July 1, 2002.

Juma, C. 2000. Science, Technology and Economic Growth: Africa's Biopolicy Agenda in the 21st Century. Tokyo: United Nations University Press.

Kufoniyi, O. 2001. RECTAS: Blazing the Trail in Capacity Building for Geographic information Production and Management of Land, Environment and Natural Resources in Africa. Presentation to the Second Meeting of the Committee on Development Information (CODI). Addis Ababa. Available at <http://www.uneca.org/eca_resources/Conference_Reports_and_Other_Docuemtns>. Acccessed August 29, 2002.

Labatut, J.-P. 2002. IDRC and the African Development Challenges. Manuscript, Carleton University, Ottawa, Canada.

Lindley, W. I., L. Van Croweder, and N. Doron. 1996. Education in Agriculture: Links with Development in Africa. FAO Research, Extension and Training Division. Available at <http://www.fao.org/sd/EXdirect/EXan0008.htm>. Accessed July 1, 2002.

Mehta, G. 2000. Science and Technology Issues. In Universities and Development, E. Bourne, ed. London: Association of Commonwealth Universities.

Murphy, A. (ed.) 1995. Geographic Approaches to Democratization: Report of a workshop prepared for the National Science Foundation. Available by email to <jwharrin@nsf.gov>.

Mwebaza, R. 2002. The Global Access Initiative: The Uganda case study. Presentation at the National Research Council, Washington, DC. January 10-11, 2002. Available at <http://www.opengis.org/gisd/cgi/listgisdtxt.pl>. Accessed 5 June 2002.

NEPAD (New Partnership for Africa's Development). 2001. A New African Initiative: Merger of the Millennium Partnership for the African Recovery (MAP) and Omega Plan, Pretoria. NEPAD.

NRC (National Research Council). 1993. Towards a Coordinated Spatial Data Infrastructure for the Nation. Washington DC: National Academy Press.

NRC. 1999. Our Common Journey: A Transition Toward Sustainability. Washington, DC: National Academy Press.

NSIF (National Spatial Information Framework). 2000. Spatial Data Infrastructure in Africa. Available at <http://www.nsif.org.za/surveyresults.html>. Accessed July 12, 2002.

Nsouli, S. M. 2000. Capacity Building in Africa: The role of international financial institutions. Finance and Development 37(4). Available at <http://www.imf.org/external/pubs/ft/fandd/2000/12/nsouli.htm>. Accessed May 19, 2002.

Rabenoro, C. 2001. Remarks at the Interacademy Panel: Capacity Building for Academies in Africa. Trieste, Italy. May 16-18, 2001. Third World Academy of Sciences (TWAS) Newsletter 13, Trieste, Italy: TWAS.

Sass, J. and L. Ashford. 2002. Women of the World 2002. Washington, DC.: Population Reference Bureau.

Schmidheiny, S. 1998. Changing Course: A Global Business Perspective of Development and the Environment. Cambridge, MA: MIT Press.

Swartz, D. 2000. The State of Universities in the Commonwealth. In Universities and Development, E. Bourne, ed. London: Association of Commonwealth Universities.

Task Force on Higher Education and Society. 2000. Peril and Promise: Higher Education in Developing Countries. Washington, DC: World Bank.

Taylor, D. R. F. 1998. Modern Cartography, Policy Issues and the Developing Nations: Rhetoric and Reality. In Policy Issues in Modern Cartography. D. R. F. Taylor, ed. Oxford: Pergamon

UNAIDS (Joint U.N. Programme on HIV/AIDS). 2000. Report on the Global HIV/AID Epidemic. Available at: <http://www.unaids.org/epidemic_update/report/>. Accessed June 5, 2002.

UNDP (United Nations Development Programme). 1997a. Governance for Sustainable Development. Policy Paper. Available at <http://magnet.undp.org/policy.default.htm>. Accessed May 20, 2002.

UNDP 1997b. Corruption and Good Governance. New York: Bureau for Policy and Programme Support.

USAID/WRI (U.S. Agency for International Development / World Resources Institute). 2001. GIS: Supporting Environmental Planning and Management in West Africa. A Report of the Joint USAID/World Resources Institute Working Group for Africa. November 2001. Washington, DC: World Resources Institute.

World Bank. 2000. Peril and Promise: Higher Education in Developing Countries. Washington, DC: World Bank.

World Bank. 2002. AIDS Blunts Progress in Education. DevNews. Available at: <http://web.worldbank.org/WBSITE/EXTERNAL/NEWS/0,,contentMDK:20043530~menuPK:34457~pagePK:34370~piPK:34424~theSitePK:4607,00.html>. Accessed July 1, 2002.

9

Lessons Learned and Recommendations

INTRODUCTION

This chapter discusses lessons learned from the application of geographic information in Africa and presents the committee's conclusions and recommendations. The first section summarizes lessons learned from the GISD case-study areas and other examples. The remaining sections contain conclusions and recommendations in a structure that parallels the flow of the report, namely, the spatial data and telecommunications infrastructure (Chapter 4), geographic data and tools (Chapters 5, 6, and 7), and geospatial capacity-building (Chapter 8).

LESSONS LEARNED

Africa has a small but growing community of geographic data providers, processors and analysts, trainers, technicians, advocates, and data and information users (decision-makers). The community's growth is demonstrated by the more than 400 participants at the Africa-GIS conference in Nairobi in November 2001 in contrast to the 70 attendees at the first Africa-GIS conference in Tunis in 1993.[1] As the community grows its activities are becoming better coordinated. This community comprises African and international partners from NGOs, universities, private companies, and foreign governments, including the space and aid agencies that are a major source of geographic data, training, and support.

Efforts to expand the use of geographic information at national and regional levels are resulting in data and information for decision-making, technical training for students and professionals, and creating geospatial capacity. The capacity to manage and use geographic data and information is growing through continent-wide activities (e.g., EIS-Africa and ECA's regional centers) and partnerships (e.g., NOAA's RANET [Radio and Internet for the Communication of Hydro Meteorological and Climate Related Information Across Africa] project, the Miombo Network, and FAO's Africover project). Some of these activities like FEWS NET (Famine Early Warning System Network) have been in place for many years, whereas others like Uganda's ACODE (Advocates Coalition for Development and Environment), Burkina Faso's PNGIM (National Program for Environment Information Management), and LEWS (Livestock Early Warning System) in East Africa are new.

Needs-Driven Approaches and Data-Sharing

Needs-driven approaches and open data-sharing environments are common among effective applications of geographic information (e.g., the CBNRM [Community Based Natural Resource Management] program in Namibia, the continent-wide MARA [Mapping Malaria Risk in Africa] project, and the Miombo Network in southern Africa). The needs-driven approach of the CBNRM program has built credibility with field users, led to a strong feeling of ownership by rural people and field-based support staff, fostered a culture of sensitivity to community needs among technical institutions that are partners in the program, generated trust and a common vision among partners (communities, government, donors), and built a critical mass to enhance sustainability of the program. Data-sharing—facilitated by adopting standardized software, data formats, and file directory structure, and a metadata database—has resulted in cost savings.

The agricultural and natural resource management sectors are a likely primary source of demand for geographic information and related decision-support tools, as these sectors are the main users because the livelihoods of the majority of Africans depend on them. Additional demand will arise

[1]Daniel Tunstall, World Resources Institute, personal communication, May 10, 2002.

as African countries need to satisfy reporting requirements on treaties to which they are signatories.

Lesson Learned: Needs-driven as opposed to prescriptive approaches with provision of information in appropriate and usable forms are most likely to result in effective application of geographic information.

From Environmental Management to Sustainable Development

Geographic information and technologies are central to achieving a successful transition from traditional environmental and resource management practices to sustainable development because of their integrative quality (linking social, economic, and environmental data) and their place-based quality (addressing relationships among places at local, national, regional, and global scales).

A narrow focus on either economic development or environmental management can obscure the connections between environmental change and social, political, and economic activities, artificially separating environment from development. This separation can result in short-term, project-oriented data collection; single-issue development agendas (e.g., economic growth divorced from environmental and intergenerational equity considerations); and spurious attempts to make tradeoffs between inseparable dimensions of sustainable development, such as human well-being and environmental protection (NRC, 2002).

Sustainable development necessarily links people, their needs, and the impacts of their behavior over time (including patterns of population growth and consumption, cultural patterns, and political activities) to the environment and the economy. Consequently, data on human population distribution are fundamentally important to decision-makers as they address Agenda 21 issues.

Lesson Learned: Geographic information and technologies are central to the transition from traditional environmental management to sustainable development, that brings people to the fore, rightfully integrating environment and development.

Geographic Information at the Intersection of Sectors

Agenda 21 (UNCED, 1992) calls for integrated social, economic, and environmental data. There is growing recognition by decision-makers in Africa that problems at the intersection of agriculture and environmental management, climate change, and land-cover change, with their attendant social and economic consequences, will be at the forefront of the twenty-first century.

Technological advances fostering the integration of satellite imagery with other data (such as socioeconomic or health data) in GIS are opening new ways to synthesize complex and diverse geographic datasets, creating new opportunities for collaboration among natural and social scientists and decision-makers at all levels (e.g., the LEWS project, the Miombo Network, the MARA project, CBNRM, and SADC [Southern African Development Community]).

Lesson Learned: In this century many environmental problems will occur at the intersection of sectors. Geographic information technologies can assist people in tackling this integration challenge.

Good Governance

Societal capacity is built by governance[2] that promotes the relationships among individuals, organizations, and the larger society. In this way governance contributes to the development of geospatial capacity. Linkages that facilitate collaboration among academics, governmental and non-governmental actors, and the private sector are needed for the transition to sustainable development (NRC, 1999).

Human and organizational capacity to apply geographic information and technology to Agenda 21 issues cannot grow or be maintained unless rooted in a wider societal context that values the contributions of science and technology, upholds principles of openness and sharing of information, and provides incentives for change and adaptation. The development of a policy environment that supports the use of geographic information depends on the attention given to scientific and technological issues in general.

Geographic data, hardware, and software systems are increasingly sophisticated but it is really the political, social, economic, and educational institutions of a country that ultimately determine the application and use of these data and tools for decision-making. Good governance creates a climate in which geospatial capacity can grow and vice versa. Geographic information illuminates social and political problems, such as the uneven distribution of the benefits of economic development, lack of accountability of elected officials, and a burden of disease that impacts societal cohesion.

Lesson Learned: Good governance promotes geospatial capacity and vice versa. Access to integrated geographic information allows civil society to hold government accountable; and government creates policies that determine public access to information and public participation in the decision process.

---

[2]Governance is defined by the UNDP as "the exercise of political, economic, and administrative authority to manage a nation's affairs." Sound or good governance is defined as that sub-set of governance "wherein public resources and problems are managed efficiently and in response to the critical needs of society" (UNDP, 1997).

Barriers to Use of Geographic Information

There remain barriers to effective use of geographic information in Africa, including:

- technical limitations of accessibility to such data as inadequate telecommunications infrastructure, limited bandwidth, and low Internet connectivity (Chapter 4);
- administrative challenges of accessibility to data including lack of (1) familiarity on the part of government officials with requests for information, (2) efficient protocols for requesting government data (Chapter 8), (3) common data standards to promote sharing (Chapter 4), and (4) issues of copyright and distribution;
- inability to afford needed data and lack of availability of hard currency and foreign exchange in many countries (Chapter 6);
- educational and organizational limitations on access to data and technology including a poorly trained workforce, and limited private-sector demand to spur the development of geographic information and tools (Chapter 8); and
- ineffective transfer of technology to the local level where many decisions are made that impact sustainable development (Chapter 7) (NRC, 1999).

The available data often are not of sufficiently high spatial or temporal resolution to be useful for decision-support at the local level. Urban planners require regularly updated data at 1-meter spatial resolution to take into account the rapid pace of change in cities. In rural areas where the bulk of the population still live the minimum spatial resolution of value to agricultural extension workers and rural development specialists is that of the small farms. Existing coverages, as outlined in this report, are impressive at national and sub-national levels but virtually nonexistent at local scales. The problem is confounded by the fact that what data are available rarely reach the rural and urban decision-makers at the local level dealing with the day-to-day realities of sustainable development.

In addition to data-availability challenges, many decision-makers in developed and developing countries have no experience with GIS and other spatial decision-support tools, and thus do not appreciate their potential for using geographic information. Other impediments to implementation of spatial decision-support systems include the orientation of projects toward data production rather than application, lack of planning for the decision-support process, lack of communication between technicians and scientists within an organization, and lack of inclusion of university research that could drive data analysis (EIS-Africa, 2001). With limited geographic data and a limited appreciation for its value the ability of African countries to address Agenda 21 issues and to fulfill their international treaty obligations for environmental reporting is compromised.

Lesson Learned: There are several barriers to the use of geographic data to address Agenda 21 issues. The next section describes approaches to overcome some of these barriers.

CONCLUSIONS AND RECOMMENDATIONS

Enabling Frameworks

Spatial Data Infrastructures

Conclusion: There is no universally accepted framework for geographic data management in Africa. An integrated, interoperable approach will provide Africans with better access to more diversified data that can then be applied to specific questions or problems. Decision-making on Agenda 21 issues requires access to data from multiple sources, including international ones, and this is facilitated by standardization within a spatial data infrastructure (SDI) (Chapter 4). Countries can benefit economically from SDIs because of the possibility to use data many times for many applications.

Recommendation: Because of the potential benefits, developing countries should consider using a standardized SDI that is compatible with the emerging Global Spatial Data Infrastructure (GSDI). Data derived from international development programs (for example, those of USAID) should conform to the standards recommended by the GSDI. In this way data collected by these programs is rendered more useful.

Telecommunications Infrastructure

Conclusion: Sustainable development activities would be improved if a greater emphasis were placed on distributed systems that enabled access to multiple geographic datasets and linked networks of African scientists, data users, and organizations. An efficient telecommunications infrastructure facilitates accessibility, use, and dissemination of geographic data and information, and forms the backbone of any SDI. Although telecommunications infrastructures are improving, in Africa as in much of the developing world they often are inadequate to support efficient SDIs (Chapter 4). Access to geographic data through the Internet is limited, and connection costs and bandwidth are restrictive for data-sharing.

In response to these problems a range of organizations are developing and improving telecommunications infrastructure in Africa (e.g., the African Information Society, the African Development Forum, the African Telecommunications Union, the African Connection, USAID's Leland Initiative, and NOAA's RANET project).

Recommendation: The U.S. government (e.g., USAID and NOAA) should continue to assist African countries in improving telecommunications infrastructure so that large computer files containing geographic data can readily be distributed within national and global spatial data infrastructures.

Collection and Maintenance of Geographic Data and Information

Data Continuity and Technological Uncertainty

For geographic information to be useful for long-term sustainable development and natural resource management, the data source needs to be dependable into the foreseeable future. With the exception of development programs now capitalizing on satellite meteorological observations, most programs will conclude as demonstrations rather than becoming operational within African institutions or programs, in part because of cost and related uncertainty over future availability of data.

Geographic information technology is rapidly changing. Dramatic changes in architectures, configurations, and approaches to data processing and handling technologies are creating concerns about technological obsolescence. The issues of data continuity and rapid technological change are important considerations when building sustainable geographic information activities. Rapidly evolving technology makes it difficult to provide access to low-cost data analysis tools and to generate continuous datasets. Without some way to assure data continuity (NRC, 1995), investments by development organizations in training and capacity building will be less useful than they could be. Without assurances that these investments will be useful in the future, it will be more difficult for African governments to invest in their own capacity and infrastructure. Changes in data access policy, data cost, or the elimination of an observation program create uncertainties about long-term benefits of international programs to Africans.

Global Positioning Systems

Conclusion: GPS information is broadcast worldwide to virtually anyone in any country and is of great importance to the practical collection and use of fundamental geographic data for Agenda 21-related initiatives.

Recommendation: The utility of GPS information should not be reduced by reintroducing selective availability and its continuity should be guaranteed. The U.S. Department of Defense should continue to allow free access to GPS data.

National Polar-orbiting Operational Environmental Satellite System, Terra, and Landsat

There are low-cost sources of coarse and medium spatial resolution land-cover information for Africa. These come from sensors that include the Advanced Very High Resolution Radiometer (AVHRR) (1 × 1 km), the Moderate Resolution Imaging Spectroradiometer (MODIS) (250 × 250 m to 1 × 1 km), and Landsat satellite sensors (79 × 79 m to 15 × 15 m). The resultant datasets include Global Land Cover (AVHRR), TREES (Tropical Ecosystem Environment Observations by Satellite) (AVHRR), GeoCover Land Cover (Landsat), and Africover (Landsat). In addition to global land-cover mapping applications NOAA's AVHRR is a widely used source of satellite data for cloud and sea-surface temperature mapping in meteorological applications, natural resource management and early warning systems (e.g., FEWS NET, LEWS). This class of sensor flies onboard NOAA's "operational" satellites and its successor will likely continue operating until 2018 onboard the National Polar-orbiting Operational Environmental Satellite System. NASA's advanced MODIS sensor on its Terra satellite platform has a range of mapping applications including land cover, fire, and productivity.

Conclusion: Land-cover datasets and vegetation indexes are valuable resources for natural resource management and development planning in rural areas. Similar datasets and indices can be constructed in the future for change detection and many other applications as long as there is continued flow of data from AVHRR, MODIS, and Landsat (or their equivalents).

Recommendation: Until at least 2018, NASA, NOAA, and DOD should carry out their plan for the National Polar-Orbiting Environmental Satellite System to ensure that it supplies relatively coarse spatial and high temporal frequency observations (such as the AVHRR follow-on) that are necessary for a multitude of applications in Africa and elsewhere.

Recommendation: NASA and USGS should take measures to ensure that the Landsat data continuity mission(s) provides long-term continuous data, perhaps through making the Landsat program an operational system for land observations, to support sustainable development and natural resource management in Africa and elsewhere. NASA should also ensure that sensors on its Terra and Aqua satellites (e.g., MODIS, ASTER, AMSR-E) continue to provide data for meteorological and land observation applications.

These actions would address data continuity at the data source. One means of reducing uncertainty in the data stream at the downstream end is to develop databases using data from more than one source. Such multi-sensor approaches as the twinning of Landsat and SPOT satellite imagery for high-resolution land-cover change information promote flexibility for agencies that base new programs on the availability of data. This approach would benefit from close coordination and cooperation among international data providers and between data providers and donor agencies. Flexibility in the provision of hardware and software technologies will also be necessary; often programs are tightly coupled to specific, sometimes unique data processing and analysis systems. More use of open interoperable software environments, as promoted in SDIs, would enhance the flexibility and reduce the vulnerability of these programs.

Focused Geographic Data Requirements

Human Population Distribution

Conclusion: National census data provide the foundation for measuring population distribution and change at the national to local scales. The strengths of human population censuses arise from their completeness of coverage; continuity of statistics from census to census; and the detail that each census provides about population sub-groups in local areas. In the current worldwide development arena, such key issues as good governance, poverty eradication strategies, and the need to promote economic growth with social equity all require population and other demographic data at the detailed local scale that only a population census can provide. Moreover, there exists an increasing demand for disaggregated data at the sub-national level.

Despite these needs datasets on population distribution from many African censuses are incomplete, often owing to high costs and insufficient funds. Progress toward Agenda 21 goals is impeded by this lack of complete, reliable data on human population distribution.

Recommendation: USAID and the U.S. Bureau of the Census should provide financial and technical support to national census offices and bureaus in Africa to help them complete censuses, geographically reference the data, and make the data available in disaggregated form to decision-makers.

Very High Spatial Resolution Remotely Sensed Data

Conclusion: Many Agenda 21 issues concentrate on urban and suburban areas (Chapter 2). Addressing sustainability issues relating to urban and suburban land use (including ownership) and infrastructure requires very high spatial resolution ($\leq 1 \times 1$ m) remotely-sensed data. High-resolution data is costly whether obtained from satellites or airborne sensors. Although there are inexpensive options for obtaining high-quality coarse (1×1 km) and medium (30×30 m) spatial resolution land-cover datasets for parts of Africa, there is no economical method of obtaining very high spatial resolution imagery to inventory and monitor change in urban areas in Africa. Image grants would help to inventory and map the continually changing characteristics of urban infrastructures.

Recommendation: USAID should consider purchasing very high spatial resolution images (i.e., $< 1 \times 1$ m) on a regular basis (at 5 to 10 year intervals) for urban areas in Africa and donating them to African organizations to ensure continuity of the data source and change detection. The imagery might include airborne analog or digital photography or satellite-derived high-resolution imagery. The areas surveyed could be requested by African organizations on the basis of importance of problem and technical and organizational capacity to use the data. One model for this concept is the U.S. Science Data Buy (Box 5-3).

Elevation Data

Elevation (topographic) data have many uses (Table 5-2) but are often inaccurate, of limited extent, or nonexistent, owing to inaccessibility of Earth's mountain ranges, deserts, and forests. Even where access is practical, traditional surveying methods are expensive. Furthermore, neighboring countries may use differing data-collection methods that cause data discontinuities at borders, even though natural resources (e.g., rivers) often cross these borders. To address these challenges, the United States has partnered with a number of countries and organizations to produce two digital elevation datasets: the GTOPO30 (Global Topography at 30 arc seconds) dataset (with its derivative hydrologic product: HYDRO1K) and the 2000 Shuttle Radar Topography Mission (SRTM) dataset. The GTOPO30 dataset is a global digital elevation dataset, whereas the SRTM dataset covers 80 percent of the globe. In the current plan, which is not finalized, SRTM data will be released at 30×30 m spatial resolution for the U.S. and at 90×90 m spatial resolution for the rest of the world.

Conclusion: The GTOPO30 dataset is of limited value in Africa and most other developing countries for monitoring ecosystems, urban and rural infrastructures, and hydrology because of its coarse spatial resolution (1.1×1.1 km). Fortunately, elevation data derived from NASA's Shuttle Radar Topography Mission in 2000 may be more suitable for many applications because all data were collected during a single 11-day mission using standardized technology; they will have accurate geodetic control; and will be homogeneous across each continent (Chapter 5).

Recommendation: NASA should produce digital elevation data from the Shuttle Radar Topography Mission at the highest possible spatial resolution (e.g., 30×30 m) for all areas. The data should be made available without restriction and at affordable cost. NASA should also provide the synthetic aperture radar ortho-image mosaics at 30×30 m spatial resolution that are being produced as part of the processing. These mosaics would provide additional information about land-cover conditions and surface roughness characteristics, especially in tropical regions shrouded by cloud cover.

Conclusion: A valuable hydrologic product for application to Agenda 21 issues could be derived from the Shuttle Radar Topography Mission with almost global 90 × 90 m (perhaps 30 × 30 m) spatial resolution. This derivative product would have applications at the sub-regional level where the low-resolution (1.1 × 1.1 km) HYDRO1K dataset currently is inapplicable.

Recommendation: Serious consideration should be given by the USGS to modeling the Shuttle Radar Topography Mission-derived 30 × 30 m digital elevation data to produce the most accurate, affordable hydrologic network database with global coverage.

Legacy Data

The earliest baseline against which future change can be compared often comes from historical legacy data. In many instances legacy data may be digitized, placed in a GIS, and analyzed in conjunction with more recent geographic data, such as satellite remotely-sensed data. The time scales over which change can be detected are extended through use of legacy data. Additionally, they contain place names and provide valuable insights on ethnicity and population growth. Bridges between local knowledge and modern technology are built through the use of legacy data.

Despite their obvious benefits, legacy data are being lost or remain inaccessible and unused. Efforts are underway to preserve legacy data and ensure that they are used. International organizations including the French Institute of Scientific and Technological Research for Cooperative Development and DEVECOL, and African regional organizations such as the Fundamental Institute of Black Africa in Dakar, Senegal, and the University of Ibadan in Nigeria are working to preserve legacy data and make it available to decision-makers.

Recommendation: To complement these efforts to preserve and enhance the use of valuable legacy data U.S. government agencies (e.g., USAID and USGS) should assist African countries and organizations to identify, integrate, and maintain existing sources of information (legacy data). They should also provide African countries with copies of such legacy data as reports, maps, statistics, aerial and satellite photographs, and other relevant data and materials currently held outside those countries. The first task would be substantial, but the second would be more routine.

Cadastral Data

Owning land provides individuals with economic assets that can be traded in land markets, used as collateral to raise credit or as security for various forms of economic improvements. Because individual land ownership is nonexistent in large parts of rural Africa, except in eastern and southern Africa, challenges remain for rural Africans to obtain credit from lending institutions in their bid to improve quality of life.

Conclusion: The production of cadastres is costly and has been a low priority for most African countries and donor agencies, even when there are clear benefits. GPS, used in concert with GIS, produces cadastral data more cheaply than traditional surveying techniques and will facilitate production of cadastres. Continued, cautious development of cadastres will facilitate land management and administration, promote greater efficiency in the operation of land markets, strengthen the operations of free-market economies, and reinforce the ability of governments to initiate and sustain land and agrarian reforms (e.g., de Soto, 2000).

Over time cadastres could play an indirect role in poverty reduction, especially through enhancing access to credit facilities and providing socioeconomic information for effective settlement management. Many Africans have no easily located residential addresses to facilitate their effective participation in social and economic transactions (ECA, 2001). These inadequacies have been one reason why the systematic delivery, management, expansion, and improvement of services to all segments of the population, the effective collection of taxes and rates, and the cost recovery for utilities and services have been difficult to implement in urban areas (ECA, 2001).

Recommendation: Because of the potential of cadastres to address Agenda 21 issues, including poverty reduction and land resource management, the U.S. government (USAID and USGS) should assist African countries to develop cadastres.

Decision-Support Systems

Conclusion: Decision-support in the area of land cover (Chapter 6) will be one of the most fruitful applications of geographic data and tools in Africa. The livelihoods of the majority of Africans depend on agriculture and natural resources, and there are many pressing problems within these sectors. Addressing these problems demands better data and better ways of analyzing the relationship between human activities and changes on the land surface. International activities could accelerate the use of decision-support systems for land-cover applications in Africa. Strategies to improve or create these datasets are needed, and these strategies would work best when built on existing initiatives and networks.

Recommendation: An effective land-cover decision-support system should include a standard classification system; baseline data and change detection capabilities; hot spot detection and high risk zone prediction capabilities; analysis and modeling of proximate (mainly human) causes of change; linkages between direct observations, case studies, and models; and established environmental indicators.

Geospatial Capacity-Building for Sustainable Development

Coordination and Partnering for Meeting African Data Needs

Conclusion: Moving beyond the current state of the art in the application of geographic data in Africa will require greater coordination among data providers, donor agencies, and the science community and end-users in Africa. Already the requirements for the next generation of remote-sensing systems are being defined or developed, yet there appears to be little dialog between the space agencies and the donor agencies, and even less input from potential end-users of the data in Africa. Lessons learned in the application of existing data for decision-making could be fed back into the definition of future observation and data system requirements, particularly in government science agencies.

Recommendation: Data providers, U.S. government agencies, and partners should work closely with African organizations to define and integrate the data needs of Africans into future programs (e.g., for new satellite remote-sensing missions) and to maximize efficiency of new programs through a coordinated approach.

Partnerships for Capacity-Building

Conclusion: Partnerships promote sharing of resources, improved communication and cooperation, and acceptance of shared standards required for spatial data infrastructures. Effective use of geographic information science in sustainable development will be associated with the strengthening of existing partnerships and the emergence of new forms of partnerships involving universities, industry, government, and civil society. Partnerships among universities and the private sector in geospatial capacity building are key to achieving a balance between supply and demand for geographic information, tools, and services in Africa. Research networks that develop as a result of these partnerships promote broad exchange of information on sustainable development, including best practices. Development of effective partnerships requires the support and incentives of both African and international donor governments.

Recommendation: In promoting organizational cooperation, emphasis should be placed on fostering innovation and the transfer of geographic data and technology through: (1) partnerships and research networks among government agencies, research and training institutions, the private sector, and the non-governmental sector; (2) international collaboration involving developed and African countries; and (3) cooperation between African and other developing countries.

Human and Organizational Capacity

Most of the existing geographic information activities in Africa were initiated in response to humanitarian needs such as famine and natural disaster, and implemented through focused, time-limited projects. Learning to apply modern geographic information and tools to address evolving societal needs requires a long learning period and attention to the development of societal capacity in science and technology. Universities are the logical source of this kind of education and training because they focus on teaching and research. With the appropriate policies and incentives, universities are also natural incubators for enterprises and social organizations. The organizations of civil society are important in many African countries where the functions of the state are inadequate.

Conclusion: It is unlikely that long-term capacity-building in technical fields such as geographic information science can be sustained in the absence of strong foundations in higher education with emphasis on science and technology. Despite the difficulties African universities face, they remain vital to the generation of new knowledge and have the potential for organizational capacity-building. The application of geographic information to sustainable development will depend on the quality, character, and direction of university education in Africa. There is an urgent need to coordinate and strengthen the capacity of university departments providing both research and training in geographic information science.

Recommendation: African universities should become a focus for capacity-building including training and research in geographic information science and development organizations should coordinate their efforts to achieve this goal.

Conclusion: A cadre of well-trained individuals will need to be formed in each country to apply geographic data and information in support of sustainable development in Africa.

Recommendation: Continuing and on-the-job training should become an integral part of the process of enhancing geospatial capacity. Organizations that provide professional training in geographic information sciences such as regional centers and polytechnics should be strengthened.

SUMMARY

Geographic data lie at the heart of many Agenda 21 issues. These data are already in use in a growing geographic information community in Africa. Although there exist a number of barriers to effective application of geographic data to Agenda 21 issues, it is likely that demand for these data will quicken the pace toward the disappearance of these barriers.

REFERENCES

de Soto, H. 2000. The Mystery of Capital: Why Capitalism Triumphs in the West and Fails Everywhere Else. New York: Basic Books.

ECA (UN Economic Commission for Africa). 2001. The Future Orientation of Geoinformation Activities in Africa: A Position Paper. Available at <http://www.uneca.org/eca_resources/Conference_Reports_and_Other_Documents/disd/codi/docs/doc3EN.pdf>. Accessed May 15, 2002.

EIS-Africa (Environmental Information Systems—Africa). 2001. Environmental Information Systems Development in Sub-Saharan Africa: Approaches, Lessons and Challenges, E. Gavin and J. Gyamfi-Aidoo, eds. EIS-Africa.

NRC (National Research Council). 1995. Earth Observations From Space: History, Promise, and Reality. Washington, DC: National Academy Press.

NRC. 1999. Our Common Journey: A Transition Toward Sustainability. Washington, D.C.: National Academy Press.

NRC. 2002. Community and Quality of Life: Data Needs for Informed Decision Making. Washington, DC: National Academy Press.

UNCED (United Nations Conference on Environment and Development). 1992. Report of the United Nations Conference on Environment and Development, Rio de Janeiro, June 3-14, 1992. Rio de Janeiro: UN. (Annex I: Rio Declaration; Annex II: Agenda 21)

UNDP (United Nations Development Program). 1997. Governance for Sustainable Development. Policy paper. Available at <http://magnet.undp.org/policy.default.htm>. Accessed May 20, 2002.

Appendixes

A

Biographical Sketches of Committee Members

JOHN R. JENSEN, *chair,* is a Carolina Distinguished Professor in the Department of Geography at the University of South Carolina. He majored in physical geography and analytical cartography and remote-sensing at the following institutions: B.A., California State University at Fullerton, 1971; master's, Brigham Young University, 1972; Ph.D., UCLA, 1976. He has mentored 50 master's students and 22 Ph.D.'s in remote-sensing. Dr. Jensen's research has focused on: (1) remote sensing of coastal wetland biophysical resources (biomass, leaf-area-index, percent canopy closure); (2) development of improved digital image-processing algorithms to extract and model change; (3) development of error evaluation statistics for assessing the accuracy of multiple-date change detection, (4) improvement of the remote-sensing and GIS-supported coastal environmental sensitivity index (ESI) mapping used worldwide for protecting coastal resources in the event of an oil spill; and (5) modeling water quality parameters (chlorophyll, dissolved inorganic matter, suspended sediment) in estuaries and reservoirs using high spatial and spectral resolution remote sensor data. He is a past president of the American Society for Photogrammetry and Remote Sensing.

KWESI BOTCHWEY is the director of Africa research and programs at the Center for International Development at Harvard University and was the minister of finance in Ghana from 1982 to 1995. As minister of finance he was key to the implementation of one of the most far-reaching economic reform programs in sub-Saharan Africa. He holds a bachelor of law degree from the University of Ghana, a master's degree in law from Yale Law School, and a doctorate from the University of Michigan Law School. He has taught at the University of Zambia, University of Dar es Salaam, and the University of Ghana. Dr. Botchwey is a member of a panel of high-level personalities on African development set up by the U.N. Secretary-General and has served as the chairman of the Economic Committee of the Global Coalition for Africa since its inception. He also serves on a number of other important boards, including those of the African Capacity Building Foundation and the African Economic Research Consortium. He has consulted widely for a number of international institutions, including the World Bank, the International Monetary Fund, UNDP, UNCTAD, and the Commonwealth Secretariat.

ELLEN BRENNAN-GALVIN is chief of the Population Policy Section of the U.N. Population Division. Since the early 1980s, she has conducted research on urbanization and urban environmental issues in more than 20 cities throughout Asia, Africa, and Latin America and is the author of numerous case studies published by the United Nations. She is a member of the U.S. National Research Council's Committee on Population and the Panel on Urban Population Dynamics. Dr. Brennan-Galvin is a resident fellow at the Woodrow Wilson International Center for Scholars in Washington, D.C., where she will work on a project entitled "Beyond Pretty Maps: Geographic Information Technology in Urban Governance." Dr. Brennan-Galvin is a Phi Beta Kappa graduate of Smith College, holds an M.A. and a Ph.D. from Columbia University, and was a population council fellow, studying demography at the Office of Population Research, Princeton University.

CHRIS J. JOHANNSEN is professor of agronomy and director, Laboratory for Applications of Remote Sensing, Purdue University. His B.S. and M.S. degrees are from University of Nebraska and his Ph.D. is from Purdue University. Dr. Johannsen has worked on soil conservation, land use, and precision farming topics using remote-sensing and GIS and GPS technologies. He is the author or coauthor of over 185 articles, papers, and book chapters and has edited a book on remote-sensing. Dr. Johannsen is active in many professional societies, having served as international president of the Soil and Water Conservation Society (SWCS). He is a fellow of SWCS, American Society of Agronomy, Soil Science Society of America, and the American Society for Photogrammetry and Remote Sensing. He is recognized as a national and international authority on land use and agricultural applications of remote-sensing.

CALESTOUS JUMA is professor of the practice of international development and director of the Science, Technology, and Innovation program at Harvard University's Kennedy School of Government. He is a former executive secretary of the U.N. Convention on Biological Diversity and founding executive director of the African Centre for Technology Studies in Nairobi (Kenya). He is chancellor of the University of Guyana, a member of the Kenya National Academy of Sciences, fellow of the World Academy of Art and Science, fellow of the New York Academy of Sciences, and member of the U.S. National Research Council's Board on Agriculture and Natural Resources. He has won several international awards, including the Pew Scholars Award in Conservation and the Environment, the United Nations Global 500, and the Henry Shaw Medal. He holds a Ph.D. in science and technology policy studies from the science policy research unit at the University of Sussex (U.K.) and has written widely on science, technology, and sustainable development.

AKIN L. MABOGUNJE is the chairman of the Development Policy Centre, Ibadan, Nigeria. His research interests include migrations, urban and regional development, and environmental management. He was formerly professor of geography and dean of the Faculty of the Social Sciences at the University of Ibadan and former president of the International Geographical Union. He is a foreign associate of the U.S. National Academy of Sciences and a recipient of both the Nigerian National Order of Merit and Commander of the Order of the Niger.

ROBERTA BALSTAD MILLER is director of the Center for International Earth Science Information Network, Columbia University. Her research interests include the role of the social sciences in public policy in South Africa and the integration of socioeconomic and remote-sensing data. She has served as director of the Division of Social and Economic Sciences at the U.S. National Science Foundation and was founder and first executive director of COSSA, the Consortium of Social Science Associations. Dr. Miller was a senior fellow at Oxford University in 1991-92 and guest scholar at the Woodrow Wilson International Center for Scholars in 1994. She chairs the U.S. National Research Council's Steering Committee on Space Applications and Commercialization and serves on the Committee on Global Change Research.

KEVIN PRICE is a professor of geography and associate director of the Kansas Applied Remote Sensing program at the University of Kansas. He received a B.S. and M.S. from Brigham Young University in botany and range science and a Ph.D. in geography from the University of Utah. He has 225 publications, has presented 159 scientific papers, and is principal/co-principal investigator on 73 grants/contracts totaling over $21 million. He has ongoing research and educational activities in the U.S. Great Plains, Central Asia, Mexico, Central America, and south-central Africa. His research focus is on land-cover and use characterization and earth system studies using observations made from satellite remote-sensing instruments.

PRISCILLA REINING, a social anthropologist and Africanist, was formerly program director in the international office of the American Association for the Advancement of Science (AAAS). She received an A.B., A.M., and Ph.D. from the University of Chicago, and is a fellow of AAAS, the African Studies Association, American Anthropological Association, and a member of the Society of Women Geographers. She is a board member of the Renewable Natural Resources Foundation. Dr. Reining has made numerous trips to Africa for field research, and an enduring concern is the relationship between human groups and their environment.

DAVID L. SKOLE is professor of geography and director of the Basic Science and Remote Sensing Institute at Michigan State University. His research interests are focused on the role of land-use and cover change and its relation to global change and sustainability development. Much of the work involves remote-sensing at continental scales in both the tropics and temperate zones, including assessment of the rates and geographic patterns of tropical forest conversion and fragmentation. His research also incorporates geographical information and geospatial information technologies in interdisciplinary analyses of the drivers of landscape change and its effect on biodiversity and biogeochemistry of natural and managed landscapes. He is past chair of the IGBP-IHDP Core Project on Land Use and Cover Change. He currently serves as chair of the Forest Cover Characteristics and Changes Implementation team of the U.N. GTOS program on Global Observations of Land Cover Dynamics, and has served on several advisory committees to federal agencies and the aerospace and GIS industries in the United States. He is currently a member of the U.S. National Science Foundation Advisory Committee on Environmental Research and Education and a member of NASA's Landsat 7 Science Team.

ANDREW STANCIOFF is a natural resources and environmental consultant with Stone Environmental, Inc., of Montpelier, Vermont. His interests and recent research include work in climate-monitoring, poverty and vulnerability assessment, as well as disease and conflict prediction. Much of this work has been done in Africa. Mr. Stancioff spent five years managing the USAID/CILSS AGRHYMET program in Niamey, Niger. He is a member of the Society of Economic Geologists and the American Society of Photogrammetry and Remote Sensing. In early 2001 Mr. Stancioff managed a project to create an atlas of poverty and vulnerability for the government of Niger with funding from the World Bank. He recently taught a course at Georgetown University entitled Environment, Resources and Conflict.

FRASER TAYLOR is Chancellor's Professor of International Affairs and Geography and Environmental Studies at Carleton University, Ottawa, Canada. He is also director of the Geomatics and Cartographic Research Centre and the Centre for Development Research and Training at the university. He has worked extensively on African development issues and has published widely in this field. A major research interest is the application of geomatics to development problems. Dr. Taylor also has extensive publications in the field of geomatics and cartography. He is an honorary life member of the Canadian Association of African Studies and served as president of the International Cartographic Association from 1987 to 1995. He is currently president of the Canadian Association of Geosciences and History for the Americas and chairman of the International Steering Committee for Global Mapping.

NRC Staff

PAUL M. CUTLER, ***study director,*** is a program officer for the Board on Earth Sciences and Resources of the U.S. National Research Council. He received a bachelor's degree from Manchester University, England, a master's degree from the University of Toronto, and a Ph.D. from the University of Minnesota. Prior to joining the NRC Dr. Cutler was an assistant scientist and lecturer in the Department of Geology and Geophysics at the University of Wisconsin-Madison. His research is in surficial processes, specifically glaciology, hydrology, and quaternary science. In addition to numerical modeling and GIS-based research he has conducted field studies in Alaska, Antarctica, arctic Sweden, the Swiss Alps, Pakistan's Karakoram mountains, the midwestern United States, and Canadian Rockies. He is a member of the Geological Society of America, American Geophysical Union, Geological Society of Washington, and is a fellow of the Royal Geographical Society.

LISA M. VANDEMARK, ***study director,*** is a program officer for the Board on Earth Sciences and Resources of the U.S. National Research Council. She has a Ph.D. in geography from Rutgers University and an M.S. in human ecology from the University of Brussels, Belgium. Her B.S. (nursing, specialty psychiatry) is also from Rutgers University. Prior to joining the NRC she was a research associate at the Institute of Marine and Coastal Sciences, Rutgers University, and an intern at the National Science Resources Center at the Smithsonian Institution. Her research interests include environmental perception and decision-making, natural resource management, land-use decisions, and the role of interdisciplinary studies in environmental protection.

ANTHONY R. DE SOUZA is currently director of the Board on Earth Sciences and Resources at the National Research Council in Washington, D.C. Previously he was executive director of the National Geography Standards Project, secretary general of the 27th International Geographical Union Congress, editor of *National Geographic Research & Exploration,* and editor of the *Journal of Geography.* He has held positions as a professor and as a visiting teacher and scholar at the George Washington University, University of Wisconsin-Eau Claire, University of Minnesota, University of California-Berkeley, and University of Dar es Salaam in Tanzania. He has served as a member of NRC committees. He holds B.A. (honors) and Ph.D. degrees from the University of Reading in England, and has received numerous honors and awards, including the Medalla al Benito Juarez in 1992 and the Gilbert Grosvenor honors award from the Association of American Geographers in 1996. His research interests include the processes and mechanisms of economic development and human-environment relationships. He has published several books and more than 100 articles, reports, and reviews.

KRISTEN L. KRAPF is a program officer for the Board on Earth Sciences and Resources of the U.S. National Research Council. She received a bachelor's and master's degree in environmental sciences from the University of Virginia. Previously, she was director of programs at the Renewable Natural Resources Foundation in Bethesda, Maryland. She provided staff support for numerous inter-disciplinary and multi-disciplinary programs, including national meetings, roundtables on public policy, international activities, and annual achievement awards. She also participated in editing and producing the *Renewable Resources Journal.* She is a member of the Ecological Society of America and the Association of American Geographers.

EILEEN M. McTAGUE is a research assistant for the Board on Earth Sciences and Resources of the U.S. National Research Council. She holds an M.S. in environmental science from American University and a B.S. in biology from Pennsylvania State University. Ms. McTague has interned at the National Academy of Engineering, the Renewable Natural Resources Foundation, and Discovery Creek Children's Museum.

TERESIA K. WILMORE is a project assistant for the Board on Earth Sciences and Resources of the U.S. National Research Council. She holds a B.A. in business management from the University of the District of Columbia. Previously, she was a secretary in Kenya for the Kenya School of Professional Studies and Cunningham G.M. (Kenya) Ltd.

B

Oral and Written Contributors

Ari Arimi
Project Energie Domestigue
Niger

Paul Bartel
Africa Bureau
U.S. Agency for International Development

Sam Bacharach
Open GIS Consortium, Inc.

André Bassolé
EIS-Africa
Burkina Faso

Abdou Ballo
Rural Economy Institute
Mali

Moïse Ballo
Université du Mali
Mali

Sjaak J.J. Beerens
International Institute for Geo-information Science and Earth Observation
Netherlands

Barbara Best
U.S. Agency for International Development

Jean Abdias Compaoré
Programme National de Gestion de l'Information sur le Milieu
The National Council for Sustainable Development
Burkino Faso

John Cloud
Peace Studies Program
Cornell University

Carmelle J. Cote
Environmental Systems Research Institute, Inc.

Jane D'Aguanno
National Environmental Satellite Data and Information Services
International and Interagency Affairs Office
National Oceanic and Atmospheric Administration

John A. Daly
Consultant

Paul Desanker
IGBP/START Miombo Network
Department of Environmental Sciences
University of Virginia

Alexander M. DeSheribinin
Center for International Earth Science Information Network
Lamont-Doherty Earth Observatory
Columbia University

Gaoussou Diarra
Ministère de l'Equipement, de l'Aménagement du Territoire, de l'Environnement et de l'Urbanisme
Mali

Colonel Mbareck Diop
Office of the President
Senegal

Paul Drichi
National Biomass Survey
Uganda

Ron Eastman
Graduate School of Geography
Clark University

Fernando R. Echavarria
U.S. Department of State

Julie Esanu
Policy and Global Affairs Division
National Research Council

Chukwudozie Ezigbalike
United Nations Economic Commission for Africa
Ethiopia

Mamadou Fofana
United Nations Environment Program
Ivory Coast

Robert Ford
U.S. Agency for International Development

Peter H. Freeman
Peter H. Freeman and Associates
DEVECOL

Mesfin W. Gebremichael
Southern and Eastern African Mineral Centre
Tanzania

Barry N. Haack
Department of Geography and Earth Science
George Mason University

Lynne Z. Hale
Coastal Resource Center
University of Rhode Island

Anders Halvorsen
World Information Technology and Services Alliance

Doris Haywood
U.S. Department of State

David Healy
Stone Environmental, Inc.

Frank Holsmuller
Environmental Systems Research Institute, Inc.

Richard Humphries
Southern African Regional Poverty Network
The Human Sciences Research Council
South Africa

Charles Hutchinson
Office of Earth Science
National Aeronautics and Space Administration

Maman-Sani Issa
Agence Béninoise Pour l'Environnment
République du Benin

Mike Jensen
Consultant
South Africa

John A. Kelmelis
U.S. Geological Survey

Mahamadou Sekou Keita
Institut Géographique du Mali
Mali

Beth Lachman
RAND

Robert W. Lake
Center for Urban Policy Research
Rutgers University

Kate Lance
International Program
EROS Data Center
U.S. Geological Survey

John S. Latham
Food and Agriculture Organization
of the United Nations
Italy

Scott A. Loomer
Advanced Research and Development Division
National Imagery and Mapping Agency

Joseph Matere
International Livestock Research Institute
Kenya

Robert L. McCanna
International and Policy Office
National Imagery and Mapping Agency

Brian J. Mhango
University of the Witwatersrand
South Africa

Allen Z. Miller
Information Technology Association of America

Roger Mitchell
Earth Satellite Corporation

John Moeller
Federal Geographic Data Committee

Keith M. Moore
Office of International Research and Development
Virginia Tech University

G. C. Mulaku
Department of Surveying
University of Nairobi
Kenya

Rose Mwebaza
Advocates Coalition for Development and Environment
Uganda Access Initiative Case Study
Makerere University
Uganda

Kate Newman
World Wildlife Fund

Jorge A. S. Oliveira
Mitchell Group, Inc.
Mali

Wilber K. Ottichilo
Regional Center for Mapping of Resources for Development
Kenya

Malam Souley Oumarou
Programme Recensement Général de L'Agriculture
Niger

Jennifer Phillips
International Research Institute for Climate Prediction
Columbia University

Mark Reichardt
Open GIS Consortium, Inc.

Patrice Sanou
Centre SIGET
Burkino Faso

Craig Schwabe
GIS Centre
The Human Sciences Research Council
South Africa

Janice Sessing
National Oceanic and Atmospheric Administration

Sydney Simelane
Department of the Surveyor General
Swaziland

Lane Smith
U.S. Agency for International Development

Kelly Sponburg
RANET

Gregory Stuart-Hill
World Wildlife Fund/ Life Project
Namibia

Jonathan G. Tagg
World Wildlife Fund /Life Project
Namibia

Drissa Tangara
Opération Haute Vallée du Niger
Mali

Gray Tappan
International Program
EROS Data Center
U.S. Geological Survey

Assize Touré
Ecological Monitoring Center
Senegal

Amidou Traore
Secrétariat Technique Permanent du Cadre Institutionnel de la Gestion des Questions Environnementales
Mali

Pierre C. Sibiry Traore
Institut d'Economie Rurale
Crops Research Institute for Semi-Arid Tropics
Mali

Daniel B. Tunstall
World Resources Institute

James Verdin
International Program
EROS Data Center
U.S. Geological Survey

Hamdou-Rabby Wane
CERPOD/Sabel Institute (INSAH)
The Watson Institute for International Studies,
Brown University

William Wood
Bureau of Intelligence and Research
U.S. Department of State

Robert Winterbottom
Environment and Natural Resources Division
International Resources Group

Robert Zomer
International Centre for Research in Agroforestry
Kenya

C

FGDC Statement

Federal Geographic Committee Policy Statement
Support for International Infrastructure Activities

The Federal Geographic Committee (FGDC) includes as its core, a vision for a Spatial Data Infrastructure (SDI) that supports local to global interests:

> **"Current and accurate geospatial data will be readily available to contribute locally, nationally, and globally to economic growth, environmental quality and stability, and social progress" (National Spatial Data Infrastructure Vision, FGDC 1997).**

Acceleration of global SDI initiatives is a high priority of the FGDC. Therefore, the FGDC commits its resources wherever possible to share experiences and work with other countries in the development of national Spatial Data Infrastructures (SDI), and to provide additional momentum for the establishment of Regional and Global Spatial Data Infrastructures. Furthermore, the FGDC is committed to maintaining an International Activities Coordination staff position to assure continued focus and US leadership presence in global SDI initiatives.

The NSDI, regional SDI's, and the Global Spatial Data Infrastructure have common goals to promote information sharing, greater interoperability, and to enable sound decision-making at all levels. The FGDC will assist other nations where possible by sharing our experiences and resources, and will establish partnerships for the betterment of SDI efforts internationally. FGDC support will include, but will not be limited to:

- *Outreach and Education* - the FGDC will provide materials, expertise, and other support as appropriate to promote best practices and principles for a SDI, and to identify the benefits of sharing data.
- *Spatial Data Clearinghouse* - the FGDC will share materials, expertise, tools, and funding to help others implement spatial data clearinghouses in their countries.
- *Framework, Standards, and Metadata - the* FGDC will provide assistance in form of materials, training, and technical expertise to help inform, educate, and implement the framework concept, spatial data standards, and metadata standards that are essential to the promotion of data sharing and exchange.
- *Building and extending relationships* - the FGDC will work to link its industry, academic, and governmental membership with other nations as opportunities arise for collaboration

Recognizing the success of SDI implementation around the globe, the FGDC seeks to collaborate with other nations to understand best practices and technology initiatives of value to US government, private, and academic interests, and to identify opportunities for transnational development and prototype partnerships.

The FGDC will seek to enhance nation-to-nation collaboration in appropriate activities.

Dr. Mark Schaefer
for Secretary Bruce Babbitt, Chair
Federal Geographic Data Committee

Mr. John Moeller
Staff Director,
Federal Geographic Data Committee

D

Acronyms

AARSE	African Association for Remote Sensing of Environment
ACMAD	African Centre of Meteorological Forecasting
ACODE	Advocates Coalition for Development and Environment
ADDS	Africa Data Dissemination Service
ADG	Africover Database Gateway
AGRHYMET	Agriculture, Hydrology, Meteorology Center
AID	Africover Interactive Database
AIM	Africover Interpretation and Mapping System
ALC	African Learning Channel
ALIN	Arid Lands Information Network
ALIN-EA	Arid Lands Information Network-Eastern Africa
AMSR-E	The Advanced Microwave Scanning Radiometer-Earth Observing System
AOCRS	African Organization for Cartography and Remote Sensing
ASARECA	Association for Strengthening Agricultural Research in Eastern and Central Africa
ASCN	African Sustainable Cities Network
ASI	Italian Spatial Agency
ATSR	Along Track Scanning Radiometer
AVHRR	Advanced Very High Resolution Radiometer
BADEA	Bank for Economic Development in Africa
BATS	Biosphere Atmosphere Transfer Scheme
BIFAD	Board for International Food and Agriculture Development
BRGM	Bureau of Geological and Mining Research
CABGLEN	Capacity Building for Geographic Information Production and Management for Sustainable Local Environment and Natural Resources Management
CAD	Computer Aided Design
CBNRM	Community Based Natural Resource Management
CDMA	Code Division Multiple Access
CEN	European Committee for Standardization
CEOS	Committee on Earth Observation Satellites
CERPOD	Center of Applied Research in Population and Development
CGIAR	Consultative Group on International Agricultural Research
CIDA	Canadian International Development Agency
CIESIN	Center for International Earth Science Information Network
CILS	CEOS Information Locator System

CILSS	Inter-State Committee for Drought Control in the Sahel
CIUEM	Information Science Center, Eduardo Mondlane University
CNRIT	Center for Natural Resource Information Technology
CODI	Committee On Development Information
CRSP	Collaborative Research Support Program
CRTEAN	Regional Center for Remote Sensing of North African States
CRTO	Regional Center of Toronto
CSE	Ecological Monitoring Center
CSIR	Council for Science and Industrial Research
DAAC	Distributed Active Archive Center
DANIDA	Danish National Aid Program
DBMS	Database Management Systems
DEM	Digital Elevation Model
DEVECOL	Development Ecology
DfID	Department for International Development
DHS	Demographic and Health Surveys
DIAPER	Permanent Diagnostic Project
DIGEST	Digital Geographic Information Exchange Standards
DISD	Development Information Services Division
DLR	The German Center for Air and Space Travel
DMC	Drought Monitoring Center
DMSP	Defense Meteorological Satellite Program
DOQQ	Digital Orthophoto Quarter Quad
DOT	Digital Opportunities Taskforce
ECA	U.N. Economic Commission for Africa
ECOWAS	Economic Community of West African States
EIS	Environment Information Systems
EOS	Earth Observing System
EPA	Environmental Protection Agency
EROS	Earth Resources Observation Systems
ERTS	Earth Resources Technology Satellite (Later renamed Landsat)
ESMI	European Spatial Metadata Infrastructure
ESRI	Environmental Systems Research Institute
ETM+	Enhanced Thematic Mapper Plus
FAO	Food and Agriculture Organization
FDI	Foreign Direct Investments
FEWS NET	Famine Early Warning System Network
FEWS	Famine Early Warning System
FGDC	Federal Geographic Data Committee
FIPS	Federal Information Processing Standards
FPAR	Fraction of Photosynthetically Active Radiation
FSS	Federal School of Surveying
FTP	File Transfer Protocol
GAC	Global Area Coverage
GDTA	Aerospace Remote Sensing Development Group
GIS	Geographic Information System
GISciences	Geographic Information Sciences
GISD	Geographic Information for Sustainable Development
GOFC	Global Observations of Forest Cover
GOLD	Global Observations of Land Dynamics

GPS	Global Positioning System
GPW	Gridded Population of the World
GRID	Global Resource Information Database
GSDI	Global Spatial Data Infrastructure
GSI	Geographical Survey Institute
GSM	Global System for Mobile Communication
GTOS	Global Terrestrial Observing System
GTS	GNU Triangulated Surface
GTZ	German Agency for Technical Cooperation
HIPC	Highly indebted poor countries
ICA	International Cartographic Association
ICLARM	International Center for Living Aquatic Resources Management
ICLEI	International Council for Local Environment Initiatives
ICT	Information and Communications Technology
IDRC	International Development Research Center
IFREMER	French Institute for Oceanic Research
IGAD	Intergovernmental Authority on Development
IGADD	Intergovernmental Authority on Drought and Development
IGBP	International Geosphere/Biosphere Programme
IGBP-DIS	International Geosphere/Biosphere Programme Data and Information System
IGN	National Geographic Institute
IHDP	International Human Dimensions Programme on Global Environmental Change
ILRI	International Livestock Research Institute
IMS	Institute for Marine Sciences
INEGI	National Institute of Geographic and Information Science and Statistics
INRSH	Nigerian Institute of Social Science Research
INSAH	Institute of the Sahel
IRI	International Research Institute
ISO	International Standards Organization
ISS	International Space Station
IUCN	World Conservation Union
JPL	Jet Propulsion Laboratory
JRC	Joint Research Center
JSG	Joint Steering Group
KARI	Kenya Agricultural Research Institute
KEMFRI	Kenya Marine and Fisheries Research Institute
LA21	Local Agenda 21
LAI	Leaf Area Index
LAN	Local Area Network
LCCS	Land Cover Classification System
LEWS	Livestock Early Warning System
LIDAR	Light Detection and Ranging
LQI	Land Quality Indicator
LTRS	Laboratory for Remote Terrestrial Sensing
LUCC	Land Use and Cover Change

MARA	Mapping Malaria Risk in Africa
MIT-AITI	Massachusetts Institute of Technology-African Internet Technology Initiative
MODIS	Moderate Resolution Imaging Spectroradiometer
MSS	Multi-spectral Scanner
NAFGIM	National Framework for Geospatial Information Management
NAPA	National Academy of Public Administration
NARS	National Agricultural Research Centers
NASA	National Aeronautics and Space Administration
NASDA	National Space Development Agency (Japan)
NATO	North Atlantic Treaty Organization
NAVSTAR	Navigation Satellite Timing and Ranging
NDVI	Normalized Difference Vegetation Index
NEMC	National Environmental Management Council
NEPAD	New Partnership for Africa's Development
NESDIS	National Environmental Satellite, Data, and Information Service
NGDC	National Geophysical Data Center
NGO	Non-governmental Organization
NICI	National Information and Communication Infrastructure
NIIRS	National Imagery Interpretability Rating Scale
NIMA	National Imagery and Mapping Agency
NIRS	Near Infrared Spectroscopy
NNF	Namibia Nature Foundation
NOAA	National Oceanic and Atmospheric Administration
NPOESS	National Polar-orbiting Operational Environmental Satellite System
NSDI	National Spatial Data Infrastructure
NSF	National Science Foundation
NSIF	National Spatial Information Framework
NUTBAL	Nutritional Balance Analysis Model
OAU	Organization for African Unity
ODA	Official Development Assistance
ODA	Overseas Development Aid
OGC	Open GIS Consortium
OLS	Operational Linescan System
ONC	Operational Navigation Chart
ORNL	Oak Ridge National Laboratory
OSS	Observatory of the Sahara and the Sahel
PNGIM	National Program for Environment Information Management
PLSS	Public Land Survey System
PMT	Photo Multiplier Tube
POES	Polar-Orbiting Environmental Satellite
PTO	Public Telecommunications Operator
PVO	Private Voluntary Organization
RCMRD	Regional Center of Mapping Resources for Development
RCSSM	Regional Center for Services in Surveying and Mapping
RCSSMRS	Regional Center for Services in Surveying, Mapping, and Remote Sensing
RECTAS	Regional Center for Training in Aerospace Surveys
REIMP	Regional Environmental Information Project

RFE	Rainfall Estimate
RGB	Red, Green, Blue
SADC	Southern African Development Community
SADCC	Southern African Development Coordination Conference
SCAR	Scientific Committee on Antarctic Research
SCT	Spatial Characterization Tool
SDI	Spatial Data Infrastructure
SDST	Spatial Data Transfer Standards
SiB	Simple Biosphere Model
SISEI	Environmental Information System on the Internet
SPOT	Système Pour l'Observation de la Terre
SPP	Single Point Positioning
SRTM	Shuttle Radar Topography Mission
SSM/I	Special Sensor Microwave/Imager
SST	Sea-Surface Temperature
START	System for Analysis, Research, and Training
STI	Sexually Transmitted Infection
STWG	Science and Technical Working Group
TB	Tuberculosis
TCMP	Tanzania Coastal Mapping Partnership
TDMA	Time Division Multiple Access
TM	Thematic Mapper
TREES	Tropical Ecosystem Environment Observations by Satellite
TRMM	Tropical Rainfall Measuring Mission
UNDP	United Nations Development Program
UNÈCA	United Nations Economic Commission for Africa
UNECA-CODI	UNECA-Committee on Development Information
UNEP	United Nations Environment Program
UNESCO	United Nations Educational, Scientific, and Cultural Organization
UNFPA	United Nations Population Fund
UNICEF	United Nations International Children's Emergency Fund
UNITAR	United Nations Institute for Training and Research
UNL	University of Nebraska-Lincoln
URI	University of Rhode Island
USAID	United States Agency for International Development
USFS	United States Forest Service
USGS	United States Geological Survey
UTM	Universal Transverse Mercator
VSAT	Very Small Aperture Terminal
WCRP	World Climate Research Programme
WRI	World Resources Institute
WSSD	World Summit on Sustainable Development
WWF	World Wildlife Fund

E

Glossary

BANDWIDTH A measure of the amount of data that can reliably be transmitted through a channel per unit of time, typically measured in bits or bytes per second.

BIOME An ecological formation including both plants and animals. Traditionally biomes are identified in terms of their characteristic vegetation form.

CADASTRE A map that is accompanied by a register showing the ownership or possession of individual units of land to facilitate efficient land administration and expedite land market transactions.

CALLING POINTS Points from which telephone calls can be made. (A slightly different use of this term is common in computer programming.)

CAPACITY The ability to undertake certain activities, solve problems, and achieve objectives, such as interpreting a map (an example of human capacity), managing computer networks (organizational capacity), or sharing data (societal capacity). Capacity is built by enhancing these abilities.

CONNECTIVITY The degree to which a communications service is available to the public, typically measured in telephones or Internet connections per unit of population.

DECISION-MAKERS In the context of sustainable development, those who choose actions that directly or indirectly affect the environment and reside in all levels of government, the citizenry, the private sector, non-governmental organizations, or overseas development agencies, for example.

DECISION-SUPPORT SYSTEM (DSS) The integration of geographically referenced data in a problem-solving situation.

DEVELOPMENT For the purposes of this report, see sustainable development.

DEVELOPED COUNTRIES Countries with "high" gross national income as defined by the World Bank.

DEVELOPING COUNTRIES Countries with "low" or "medium" gross national income as defined by the World Bank.

DIGITAL DIVIDE This term describes the gap in access between developed and developing countries or within a country to information and communication technologies.

DIGITAL ELEVATION MODEL (DEM) A digital representation of the elevation of locations on Earth's surface. A DEM is often used in reference to a set of elevation values representing the elevations at points in a rectangular grid.

EL NIÑO An episodic global weather phenomenon driven by conditions in the western Pacific Ocean.

E-READINESS This term is used to suggest the readiness of an organization, community, or country to use the Internet generally or to use it for business, commercial, or governmental purposes. A number of organizations have proposed instruments to measure e-readiness, and these instruments differ somewhat.

GEODETIC CONTROL Reference points on Earth whose exact positions are determined to a high degree of accuracy through measurement.

GEOGRAPHICALLY REFERENCED DATA Data with known latitude, longitude, and elevation, or other horizontal and vertical coordinates.

GEOGRAPHIC INFORMATION SYSTEM (GIS) A digital information system that is designed to work with data referenced by spatial or geographic coordinates.

GEOGRAPHIC POSITION Refers to horizontal coordinates as well as elevation of objects and features.

GEOGRAPHY An integrative discipline that brings together the physical and human dimensions of the world in the study of people, places, and environments. Its subject matter is Earth's surface and the processes that shape it, the relationships between people and environments, and the connections between people and places.

GLOBAL POSITIONING SYSTEM (GPS) A network of satellites controlled by the U.S. Department of Defense that is designed to help aerial and ground-based units with an appropriate receiver determine their current location.

GSM An acronym for Global System of Mobile Communication, one of several competing standards for mobile telephones.

INFORMATION Data that humans assimilate and evaluate to solve a problem or to make a decision.

LEGACY DATA Maps, aerial photographs, reports, and other documents that often function as baseline information.

LOCAL AREA NETWORK Personal computers linked into a network in a limited geographic area, such as a building or campus, as distinguished from a wide area network or stand-alone computers.

METADATA A term used to describe information about data. Metadata usually includes information on data quality, currency, lineage, ownership, and feature classification.

OPEN SOURCE SOFTWARE Software for which the original or source code is freely available to the user or other interested parties. Most commercial software firms make the original source code available only under specific circumstances, usually to collaborating firms, providing the user with only a machine readable version of the software that is difficult to read and/or modify. In some cases, most notably Linux, an open source operating system, large communities have formed around the development and maintenance of specific open source software packages. Open source software is of special interest to developing countries in that its development and maintenance cost is born by communities of volunteers, and it is available without charge, or with minimal charges for the provision of disks, manuals, and support.

ORTHO-IMAGE A specially processed image prepared from an aerial photograph or remotely sensed image that has the metric qualities of a traditional line map with the rich detail of an aerial image.

PENETRATION The degree to which a technology has penetrated a community or country. Indicators of penetration include the number of points of presence (i.e., points at which calls are routed for transmission by a long-distance, interexchange carrier), the percentage of organizations of a given type (e.g. businesses, schools, health centers) that have access to a service or the number of secondary cities that have Internet service providers in a country.

PHOTOGRAMMETRY The art, science, and technology of obtaining reliable information about physical objects and the environment through the processes of recording, measuring, and interpreting photographic images and patterns of electromagnetic radiant energy and other phenomena.

PLACE-BASED STUDIES The systematic analysis of social, economic, political, and environmental processes operating in a place that provides an integrated understanding of its distinctiveness or character.

RADAR INTERFEROMETRY A technique in which two radar images are taken from slightly different locations and differences between these images allow for the calculation of surface elevation.

REMOTE-SENSING The measurement or acquisition of information of some property of an object or phenomenon, by a recording device that is not in physical or intimate contact with the object or phenomenon under study.

RESOLUTION A way of detecting variation. In remote-sensing, there is spatial resolution (the variation caused by distance separating adjacent pixels), spectral resolution (the variation caused by the spectral responses within a wavelength band), and temporal resolution (the variation caused by time over the same location).

RIFT VALLEY FEVER (RVF) An acute, fever-causing viral disease that affects domestic animals (such as cattle, buffalo, sheep, goats, and camels) and humans. RVF is most commonly associated with mosquito-borne epidemics during years of heavy rainfall. RVF is generally found in regions of eastern and southern Africa where sheep and cattle are raised. RVF virus also exists in most countries of sub-Saharan Africa and Madagascar.

SCHISTOSOMIASIS One of the major communicable diseases of public health and socio-economic importance in the developing world. Direct mortality is relatively low, but the disease burden is high in terms of chronic pathology and disability. The distribution is particularly related to large-scale water development. Despite control efforts in a number of countries, still an estimated 200 million people are infected, of which 120 million are symptomatic and 20 million have severe disease. An estimated 80 percent of all cases and all of the most severely affected are concentrated in Africa.

SELECTIVE AVAILABILITY The restriction of access to or the adjustment of the precision of geographic positioning information by introducing a bias to the system.

SPATIAL DATA INFRASTRUCTURE (SDI) The institutional framework, policies, technologies, and data requirements to enable spatial data to be used to support economic growth and social and environmental interests.

SUSTAINABLE DEVELOPMENT Development that meets the needs of the present without compromising the ability of future generations to meet their own needs (from *Our Common Future,* Oxford University Press, 1987).

TELECENTER A center offering telephone and other communications services to the public. There are many different kinds of telecenters and many business models under which telecenters are operated in Africa.

TELEDENSITY The number of telephone connections per unit of population, typically measured by the number of fixed lines per unit of population, number of mobile phones per unit of population, or both.

TELEPHONY The art or practice of operating telephones or telephone services.

VSAT (Very Small Aperture Terminal) Direct link to communications satellites, typically using antennas less than 2 meters in diameter. In Africa an increasing number of organizations have Internet linkages through VSAT terminals, bypassing the local telephone service.